EDISON VS. TESLA

EDISON VS. TESLA

THE BATTLE OVER
THEIR LAST INVENTION

BY

JOEL MARTIN AND WILLIAM BIRNES

Skyhorse Publishing

Skyhorse Publishing books may be purchased in bulk at special discounts for sales promotion, corporate gifts, fund-raising, or educational purposes. Special editions can also be created to specifications. For details, contact the Special Sales Department, Skyhorse Publishing, 307 West 36th Street, 11th Floor, New York, NY 10018 or info@skyhorse-publishing.com.

Skyhorse® and Skyhorse Publishing® are registered trademarks of Skyhorse Publishing, Inc.®, a Delaware corporation.

Visit our website at www.skyhorsepublishing.com.

10 9 8 7 6 5 4 3 2 1

Library of Congress Cataloging-in-Publication Data is available on file.

Cover design by Rain Saukas
Cover photo credits: Library of Congress

Print ISBN: 978-1-5107-1876-0
Ebook ISBN: 978-1-15107-1877-7

Printed in the United States of America

Dedication

This book is dedicated to Kris and to Nancy Birnes

Acknowledgments

Bill Birnes, With Much Appreciation

—Kris Rus
Catherine Erdelyi
Thomas Santorelli

Special thanks to: Leonard DeGraff, Roxanne Salch Kaplan, Patricia Ippolito, Gaylon Emerzian. Also: Cambria Weintraub, Caleb Weintraub, Arlene And Michael Rosich. Eternally: Stephen Kaplan, Chris Martin, Evelyn Moleta, John Blake, Vladimir Rus, Father John Papallo.

We are grateful to Jay Cassell, Veronica Alvarado, and to the management and editorial staff at Skyhorse Publishing for guiding this book through all of its stages.

"Acta fabula est, plaudit"

—Caesar Augustus

Contents

Foreword
By Paul H. Smith, PhD

Author, *Reading the Enemy's Mind: Inside Star Gate–America's Psychic Espionage Program* and *The Essential Guide to Remote Viewing: The Secret Military Remote Perception Skill Anyone Can Learn.*

SPIRIT PHONE

We humans are an odd bunch. We are willing to believe only what we want to, and disbelieve what we will—even in the face of compelling evidence to the contrary. As the authority of religion erodes in our modern era, scientists ironically have by default become society's presumptive standard-setters in matters of truth and fact. Yet when it comes to matters outside of their comfort zone, they often score no higher on the objectivity meter than the average scientifically illiterate person.

This is typical of human psychology—merely being a scientist gives you no immunity. We part only reluctantly with our cherished beliefs, comfortable philosophies, or well-worn ways of doing or looking at things in the world. Such was the case with the final disposition of Thomas Edison's "Spirit Telephone."

The story that here unfolds is one of genius, drama, inspiration, insight, and single-minded drive. Yet it is all about one man. If the term "larger-than-life" wasn't coined for Thomas Edison, it should have been.

Joel Martin and Bill Birnes have wisely chosen in this book to give us the framework first, to help us understand the man who was the author of this remarkable invention. It is a story worth hearing—not just for its drama, but for its inspiring qualities—and because so few of us these days know Edison's saga. We learn of his early precociousness, his missteps, his family milieu. But we also see the promise of the budding engineer, inventor, and entrepreneur—an innovator who was not just the harbinger of a new age, but its creator as well.

Any one of Edison's inventions would have been an epochal achievement, yet he had many. We learn of his idiosyncrasies, and his sometimes mean-spiritedness. But we learn more significantly of his admirable qualities—a man with truly astounding vision and insight, with an ability to not just overcome failure, but to integrate it into his creative process—embracing failure as a necessary precursor to progress.

As this book makes clear, Edison led humanity kicking and screaming from one world-changing technological revolution to the next. The battle was always uphill, and he always came away the victor—except, that is, for his last fight, his spirit telephone.

The idea and theory behind this unique communication device were every bit as ingenious as anything else Edison invented. Much of the charm and fascination of this book is the journey on which the authors invite us as they follow their sleuthing path to its conclusion. That detective work was not easy, by the way, since much (though thankfully not all) of the evidence for the device was apparently intentionally suppressed after Edison's passing.

It was as if, after all the brilliant accomplishments and the genius of his inventions, science finally said to Edison, "Communicating with spirits? Now you've gone too far." Such is typical of an attitude that seems inextricably embedded in the mindset of modern science. It is fine to let your imagination roam, explore phenomena at will, delve into the secrets of nature whether small or great—but only up to a point. There are boundaries beyond which science forbids its researchers to go. Despite long and voluminous evidence in support of it, extrasensory perception (ESP) is one of them. The possibility that consciousness in whatever form survives death is yet another, as Edison found out in the end. One might almost come to suspect that science wants death to be the end of it for each of us.

In the matter of Edison and his remarkable spirit phone, it seems that the "scientific intolerance brush" painted all parties concerned. Edison was splattered by it because, while he was willing to entertain the idea that spirits are real, he couldn't bring himself to believe they were anything more than some electrical manifestation of an otherwise strictly physical world. But he at least was willing to invest his waning energy in trying to communicate with the spirits in whatever form he believed they existed.

His scientific critics got the most liberal daubing of that paint for not being willing to entertain any theory that violated their notions of scientific propriety. And Edison's heirs and estate couldn't escape the brush either, as they attempted to expunge the Great Man's record of any hint of that kind of extraordinary claim, so his post-mortem reputation would remain clear of any perceived tarnish.

It will not be much of a spoiler to tell you that, as the end of the book approaches, the spirit telephone turns out not as a failure, exactly, but certainly not as a success—definitely not the success that accompanied Edison's other storied inventions. You know that because, just as we have universal access to artificial light, recorded music, and movies (all Edison inventions), if the spirit phone had succeeded, all of today's households would have access to a spirit phone were there ever a need—and they don't.

The spirit phone didn't work, yet I'm not quite willing to call it a failure. Why not? Because maybe it didn't fail—maybe it just didn't have a chance to succeed the way the phonograph did, or the light bulb did after its long string of failures before the final success. Maybe the elderly Edison just ran out of time. Had he been granted only a few more short years, we might even now be communicating with departed loved ones—or even Edison himself. If we could, perhaps Edison would still be adding new inventions to his already illustrious string of accomplishments thanks to his very own spirit communication device.

Preface: Tesla's Voices of the Aether
By Tim R. Swartz

Nikola Tesla was a man of science. Books written after his death speculate that Tesla's extraordinary scientific abilities were the result of psychic powers and the paranormal. Tesla no doubt would be chagrined to hear these fantastic theories, as he had no time for the idea that the paranormal was something beyond scientific understanding.

According to Tesla in his secret journals, "Physics extends beyond what is scientifically known today. The future will show that what we now call occult or the supernatural is based on a science not yet developed, but whose first infant steps are being taken as we speak!"

Unlike many scientists, Tesla was not afraid to conduct research on something considered outlandish. If his interest were roused, Tesla would devote a tremendous amount of time in an attempt to figure it out. This is how Tesla may have become involved in what is now known as Electronic Voice Phenomena . . . something unheard of at the time.

In 1898, Nikola Tesla built a laboratory near Pike's Peak, Colorado, conducting research on thunderstorms and lightning. In his lab, Tesla constructed receivers in order to use radio frequencies to indicate approaching storms. Along with the static created by lightning, Tesla became aware that his receivers were also picking up signals that were entirely unknown to him.

In an article written for *Collier's Weekly* in 1901, Tesla detailed the amazement he felt when it dawned on him that he was hearing something "possibly of incalculable consequences to mankind."

"My first observations positively terrified me, as there was present in them something mysterious, not to say supernatural, and I was alone in my laboratory at night; but at that time the idea of these disturbances being intelligently controlled signals did not yet present itself to me."

Tesla noted that he was already familiar with such electrical disturbances produced by the sun, Aurora Borealis, and earth currents. However, the electromagnetic pulses that he was hearing were something completely unknown to him:

> The thought flashed upon my mind that the disturbances I had observed might be due to intelligent control. Although I could not decipher their meaning, it was impossible for me to think of them as having been accidental. The feeling is constantly growing on me that I had been the first to hear the greeting of one planet to another.

Afterwards, other work consumed Tesla's time. But his interest in the unknown signals persisted, and he strived to perfect his receivers so they could be tuned to any electromagnetic frequency. His diligence finally paid off when Tesla began to receive human voices, at a time when radio transmitters were still practically nonexistent.

Unfortunately, there is little else known about Tesla's research along these lines. His interests took him into other directions, and no other notes have been found. It would be interesting to know what Tesla's ideas were concerning the strange voices. Did he feel they were interplanetary, or something else?

It wasn't long after that other people started hearing unknown voices over various electronic devices, voices that claimed to be the spirits of the dead. It will never be known for certain, but was Nikola Tesla the first to open the door to what later became known as Electronic Voice Phenomena?

EDISON VS. TESLA

Introduction: Why This Book?

There's no doubt that Thomas Edison was one of the most practical and hardheaded inventors of the nineteenth and twentieth centuries. In fact, more than one science historian and cultural scholar has said that Edison was instrumental in inventing technology critical to the twentieth century and creating the technology that is driving the twenty-first century. Our world today has come about because of the world Edison envisioned all the way back in the 1890s. Not only did he bring electric light to cities around the world, create the first municipal power supply grid, create the technology for motion pictures, and make it possible for musical entertainers to preserve their art for generations on phonograph cylinders and then discs, he also invented the modern technological industry and its research and development departments. Edison actually created the creation mechanism for driving new ideas and technologies. His achievement cannot be overstated.

We also know that, as an electrical engineer and an inventor, Edison was the consummate materialist, believing that all things that exist, from the blood that courses through our veins to signals speeding through wires and even thoughts that travel through the neural circuits in our brains, are composed of palpable, definable,

1

and quantifiable substances. Therefore, the authors ask, why is it that a man who was so accomplished in all things technological, so rational in his approach to utilizing the scientific method as a creative process, and so dogged in his determination to bring science to the masses, turned his attention to communicating with the dead in the final decade of his life? On the surface, Edison's system of beliefs and his determination to prove that life existed after death would seem to be an absolute contradiction. And that is what most people, particularly the skeptics among us, believe. But there is another answer, more scientifically grounded, which we explore in this book.

Edison scholars and historians have long speculated about his last invention, the spirit phone or ghost phone which, actually, was neither a phone nor something intended to communicate linguistically with ghosts. What drove him to seek the assistance of mediums and channelers, those whom he referred to in his journal as "charlatans," to help him in his quest to prove that spirits of the departed float among us? Was it something of a psychological contradiction, the fantasies of an old man deluding himself into thinking that he could accomplish a feat that oracles and shamans had been claiming they could do since the beginning of human civilization? Or was Edison really onto something? Did he figure out an experiment, notwithstanding his reliance upon clairvoyants, which might prove definitively that there is an essence of life after the body dies? And if the latter is true, then how did that comport with the cultural and intellectual trends that might have influenced Edison's view of the world?

These questions and others are addressed in the following chapters because we, like the Edison historians and scholars, were just as perplexed about Edison's fascination with a device to communicate with the dead and his demonstration of such a device to his peers in science and engineering. We believe that by looking not only at the life of Thomas Edison, but at the cultural and intellectual trends of his life, his ongoing rivalry and very personal feud with Nikola Tesla, and his own understanding of the work of great physicists such as Einstein and Planck, that the answers to the question of why a spirit phone become abundantly clear.

We begin our study with a look at Edison's childhood in one of the most exciting times in American history: the beginning of the Age of Invention, when electricity, chemistry, and industrial manufacturing—all driven by the necessities of war—combined to create an environment of discovery. Edison came of age during this environment of discovery, participated in it, and helped shape it. We look at his early education to figure out why a young man so curious about the world around him was so difficult a student. Homeschooled, Edison learned on his own, tinkering with his chemistry set in his basement, starting first a small retail and then a printing and publishing enterprise on the Grand Trunk Railroad, and becoming a telegraph operator so inquisitive that he was able to improve the device he learned on. While many biographies of Edison cover his youth and education, we approach it from the perspective that looks at Edison's lifelong role as a marketer of invention, the creator of consumer-driven devices that responded to needs in the marketplace. We show that Edison helped create and shape the Age of Invention and the marketing of consumer technology. Decades before modern office technology companies created the concept of the automated office and sold their products into it, there was Edison creating the modern electrical communications industry. He was not just ahead of his time, he created his time.

In describing the confluence of the Great Age of Spiritualism in the latter nineteenth century as well as the Age of Science and Industrialism at the beginning of the twentieth century, we approach these trends not just as historical events, but as market influences upon Edison. We show that Edison was so affected by their coming together that he sought to amalgamate them into an invention that would respond to a market he believed existed in the 1920s. Hence, our focus on what took place in the nineteenth and early twentieth centuries is to set the stage for Edison's thinking and response to what he saw happening all around him.

Why do we spend so much time with Nikola Tesla and his rivalry with Edison in this book? We do this because we argue that Edison was not only market driven, but rivalry driven. Tesla, for all of his genius and faults, was Edison's foil and his factor, inspiring him to go places where one might not expect him to go. Tesla, who created robotics, wireless transmission of electricity, alternating-current electric generators, solar-powered generators, and even a crude understanding of artificial intelligence, was not only Edison's intellectual rival, but his business rival as well. He spurred Edison to do the one thing that prophets and soothsayers had been attempting to do since human beings first crawled down from treetops and into caves: communicate with the dead. It was a challenge to Edison's vision of science and, hence, our mission to explain.

In this book, we cover Thomas Edison as a youth, running from his birth in 1847 to the beginning of his career. We then move on to the great intellectual and cultural movements that influenced him, from the Great Age of Spiritualism to Age of Science and Industrialism. Next comes his rivalry with Tesla, with spurred Edison to compete with the inventor of alternating current. Then, finally, we come to the spirit phone itself, the science behind it, what happened to the device, its legacy, and ultimately the legacy of Edison himself.

We also demonstrate that Edison's fascination with the science of his day, especially Einstein's special theory of relativity and his general theory of relativity, are borne out even in today's news. Simply stated, special relativity expresses the relationship between energy and mass without considering the effects of gravity, while general relativity accounts for the relationship among energy, mass, and gravity. Hence special relativity is a subset of general relativity.

We show that Edison was influenced by Einstein's predictions about the universe, about devices that could register the presence of unseen forces, of theories about the unity of space and time, and how remnants of events in times past linger in our reality. We need no further evidence of this than the June 2016 discovery of a gravitational wave resulting from the collision of two black holes, a wave that traveled across the universe and across 1.3 billion years. It was finally measured by a device that in some ways imitates Edison's theories about the presence of particles or wave forms that we cannot see but can register on a meter.

We look not only to the past in this story of Edison's last invention, but to the future. We suggest that the scientific promise held out by Edison as well as Tesla, their shared theory that there was a science to the paranormal, a materialism underlying the spiritualism, a physics of immortality, helped create a future that surrounds us today. Therefore, we talk about such things as Remote Viewing, now accepted operations in the military, law enforcement, and corporate spying as well as in entertainment circles. And we suggest that Edison's spirit phone, his dogged attempt to bring scientific materialism and spiritualism together at the end of his life, wasn't so strange at all. It was simply an extension of what he had been pursuing since he was a young boy tinkering with chemicals in his father's basement, pushing the envelope of discovery. Therefore, even though his business associates, his family, and other people might have thought it strange, the spirit phone was Edison's last frontier, a frontier that inspired physicists for the next century as they tried to unlock the mysteries of how quantum mechanics might be the key to creating another invention that would change the world as we understand it.

CHAPTER 1

What Was the Spirit Phone?

On a chill winter night in 1920, according to an account in the October 1933 issue of *Modern Mechanix* magazine, with the wind whistling through the darkness outside of his Menlo Park, New Jersey, laboratory, the great inventor, industrialist, and founder of General Electric, Thomas Edison, gathered a group of his scientist friends to bear witness to his latest experiment. It was a creation his guests believed uncharacteristic of him, yet in reality was completely in keeping with his scientific beliefs. He had been working on it in secret, a project that seemed contrary to the technology-driven science Edison had embraced since his childhood in Port Huron, Michigan. But this demonstration was no attempt at mediumship or channeling, though mediums were in attendance. As the gathered scientists watched, they first heard the soft hum of an electric current, then saw a glow of light from an apparatus on the workbench that looked like a motion picture projector shoot a narrow beam of light into a photoelectric cell. Edison explained that the light on the cell, like the fog in a vacuum bell jar, would register any disturbance to the continuity of the beam when any object, no matter how evanescent or ephemeral, crossed through it. The resulting registration of an object's presence would be displayed on a meter wired to the photoelectric cell, a telltale sign to the machine's operator that something was there even if invisible to the naked eye. What was the great Edison looking for, the scientists might have asked themselves? What could be crossing the beam?

Although the arrangement was different from what they'd seen before at the laboratory, the invention was made of familiar component parts. Edison's motion picture projector box had been in operation for over a decade and by 1920 was already the basis of an entirely new industry. And a photoelectric cell was a standard piece of equipment to register a beam of photons. But why would the old man project a beam of light onto a cell instead of a screen? What was this device supposed to do? Edison was cryptic toward his guests at first. But there were others present in the laboratory that night, people who were as much an anathema to the scientists as heretics were to clerics. They were the very folks Edison had dismissed as charlatans.

Along with the scientists in the room that night, participating in the demonstration of Edison's machine, were spiritualists, mediums, and channelers who used objects like Ouija boards and tea leaves to divine what they said the souls of the departed communicated to them. Edison argued that most spiritualists were fakes and didn't believe in their talismans of foretelling the future, but tonight he was making an exception. Tonight he needed them for the very thing they asserted they could do: connect with the spirits of the departed. He needed them to endorse the concept of his device.

Although Edison publicly and in his private writings had professed himself to be the consummate materialist, who marshaled the flow of electrons through circuits to provide light, record sound, and make photographic images dance across a screen, on this night he hoped to show that materialism could also explain spiritualism. He sought to combine technology with spiritualism to see if individuals who claimed to have the power to summon the departed could actually do so, and in so doing, lure the spirits they invoked across the beam of photons so as to register on an electric meter. For those in the room watching the experiment, how many of them would realize that this was a once-in-a-lifetime event, the industrialist of his age relying on the efforts of those he had once called frauds?

Thus, the pseudo-séance began, the ceremonial invocation of the spirits of the departed. Each spiritualist performed his or her own ritual to reach what he or she thought was the other side, but what Edison believed was simply still the here and now, a collection of entangled submicroscopic particles of life, units that made up the spirit, but not the flesh, of departed human beings that floated through the aether.

As the séance wore on, the scientists kept their stare fixed upon the needle on the meter, ostensibly to note the passage of an entity across the beam, waiting to see if something would register, waiting to see if there were spirits present. Were the powers of the spiritualists performing their rituals strong enough to lure discarnate units of life across the beam? Was Edison's device sensitive enough to register any interruption of the beam? Were spirits even present? They would all soon see.

Whatever the outcome of the experiment that night, Edison himself would not be deterred from his belief that some form of a spirit existed after the death of the body. Accustomed to failures of his other inventions during theory-development stages, Edison believed that each failure was actually a form of success because it eliminated a possibility. And whatever was left after he had eliminated his failed attempts would be bound to succeed. Insofar as this projection apparatus was concerned, Edison had faith that he would ultimately succeed because he believed in the current theories of physics and the presence of unseen particles imparting life to all creatures on Earth. As it turns out, he did not have enough time, as he passed away on October 18, 1931. Even on his deathbed, he was able to arouse from a coma and tell those at his side that he had indeed found there is life after death, because he had seen the other side with his own eyes. And he knew, even as he lapsed back into a coma and breathed his last, that he had been correct all along.

Thomas Edison was a scientific skeptic. Coming of age during the Great Age of Spiritualism and thence into the Age of Science and Industrialism, Edison dismissed what he called "mediumship" as a form of charlatanism. He mocked attempts to communicate with the spirit world through Ouija boards, commenting on the ridiculousness of thinking that a disembodied spirit could manifest its thoughts through a piece of wood. In his diary, he wrote: "The thing which first struck me was the absurdity of expecting 'spirits' to waste their time operating such cumbersome unscientific media as tables, chairs, and the Ouija board with its letters."[1]

Yet, being a scientist, Edison was working in an age when huge discoveries were being made in the sciences of physics, chemistry, spectrographic analysis, his own electricity, and the new disciplines of genetics, epigenetics, and biology. It was an age when Darwin had figured out how biological species evolved, when Einstein

---

1 Edison, Thomas, *The Diary and Sundry Observations of Thomas Alva Edison*, (New York: Philosophical Library, 1948), 205.

hypothesized about the nature of matter and its relationship to energy, when Max Planck was theorizing about the mysteries of quantum physics, when geneticists were formulating the theory that genes or life units imparted traits to the human embryo at fertilization that would define it in life, and when medical doctors like Freud and Jung were explaining how unseen forces operate on the human mind. What an exciting time to be alive. What an exciting time to be a part of the very community that was making these discoveries. Yet Edison felt there was more to be discovered. There was much more to understand not so much about life, but about the essence of reality that comprised life and what happened when life itself seemed to stop after a person passed away. Did that person really pass away, Edison asked, or did that person simply become translated into another form that we couldn't perceive through our usual five senses? If so, could the great inventor and scientist of his age find a way to enable that form of perception?

In an interview in the October 1920 issue of *The American Magazine*, Edison confirmed that his scientific curiosity was intrigued by the nature of what happens to us after death. In his interview he posed the question, does our consciousness simply disappear as our bodies decompose or does some essence of our personality still linger in some form in this dimension of reality? Edison admitted that he didn't know, but the scientist in him wanted to find out whether that question could be answered. He told his interviewer that he was actively pursuing a device that would help him find that answer, saying, "I have been at work for some time building an apparatus to see if it is possible for personalities which have left this earth to communicate with us." For Edison, who took his inventions very seriously, this was not a simple throwaway remark, it was an announcement.

It was, nevertheless, a cryptic statement. What did Edison mean when he said he was "at work building"? Clearly he was in a construction phase because he had demonstrated a prototype of the device. Surely, according to the details of his demonstration reported in *Modern Mechanix,* he had worked out the design of the electronics. Yet, in another interview, this time in the far more skeptical *Scientific American,* Edison qualified his remarks by telling his interviewer that, "I have been thinking for some time of a machine or apparatus which could be operated by personalities which have passed on to another existence or sphere."[2] He seemed to be saying that the control of the device would be by those who had departed. They could choose to register across the beam or not, which would have been revolutionary. And was "thinking" what Edison meant when he said he was at work "building"? Seemingly contradictory remarks like these sparked interest among the public in the 1920s because Edison, his inventions, and his pronouncements were of great import to Americans then enjoying the booming stock market and the unbridled financial optimism of the period, set against a general sense of disillusionment in the wake of the Great War.

At his core, Edison was a pragmatic inventor, creating apparatuses that he saw as satisfying consumer needs. Sometimes, the machines he invented created their own consumer markets. Wax recording cylinders, for example, fascinated the public because for the first time the human voice could be preserved; hence, the recording industry was born. When Edison perfected a camera that could capture still photographs in succession and play them back so that they displayed movement from a sequence of frames, the

---

2 Lescarboura, Austin C. "Edison's Views on Life and Death," *Scientific American* 123 (18), October 30, 1920.

motion picture industry was born. By 1920, both industries were flourishing and Edison, although a national hero, was looking for a renewed burst of relevance. And he found it amidst the merging of the Great Age of Spiritualism and Age of Science and Industrialism. In a career that boasted "firsts"—first motion pictures, first electric light bulbs, first portable recording devices, first industrial power grid—Edison now sought another first: the first scientific approach to communication with discarnate entities. What had originally been the province of faith-based religions, with pre-Christian rituals, and then the practices of trance mediums and channelers, in Edison's vision would soon be subject to scientific scrutiny, ultimately allowing for a kind of dialogue of yes and no in response to questions from the those in attendance, a form of binary code that would allow, for the first time, for electronic communication with the dead. That was the premise of the spirit phone or what popular news articles called the "Ghost Machine."

We know this device existed and that Edison was working on it because he described it and the science behind it in his diary, even though later editors of that diary excised or redacted that particular chapter. Why? We will answer that question later, but suffice it to say that in Edison's original thinking, he had set forth not only his intention to create a spirit phone or ghost machine, but provided the scientific theory behind it. This was an early twentieth-century theory that has since been borne out by the modern theoretical physics of quantum entanglement, spooky action at a distance, and, of course, Einstein's special theory of relativity.

The important thing to remember about Edison and his approach to the unconventional was that he relied on the underlying science of that approach. He wrote that although he could not be sure whether human consciousness existed on another plane after the death of the body, he believed that if Einstein's theory of special relativity were correct, then nothing really faded out of existence. Mass became energy and that energy, if coalesced into a bundle or packet, might be reachable by another packet, this one a stream of photons. Hence, the idea for a spirit phone had taken shape in his mind: photons to electrons, waves of energy converted into patterns of electric charges. And streams of electrons were the very things that Edison had been experimenting with for the previous thirty years, ever since Alexander Graham Bell perfected a telephone.

Because he believed that, on the other side of life, the energy amassed by the departed exerted a pressure upon our reality, Edison's device would act as a "valve," which is what he called it, that would not only open to allow waves of energy through it, but, he wrote in his diary, would augment that energy "in exactly the same way that a megaphone increases many times the volume and carrying power of the human voice." If we can envision the early phonograph, Edison used that same principle of a huge megaphone to increase the volume and projection of the human voice. In this way, he believed, even the minutest amount of energy, if it existed on the other side, would be amplified to a sound that human beings on this side could hear.

Edison's belief that there was a universe of eternal matter, neither created or destroyed, from which life on Earth was purely a manifestation, also had its roots in the Platonic theory of *noumena*, a world of forms, and a Jungian theory of a collective unconscious, a shared reality in which all life participates. Hence, a device that can so communicate with that collective unconscious reservoir might be able to carry signals—a flow of electrons—from that side to ours.

In addition to Edison's theory of the transportation of electrons from one side of reality to another, there was also a biological component of his theory that it might be possible to communicate with the departed. In his diary, Edison wrote of the memory storage function of an area of the brain, which we now know is situated on the left hemisphere, called "Broca's area." Edison believed this was the seat of memory, whether conscious or unconscious, that could play back images from an individual's life. We know now, however, that the seat of long-term memory is situated in the amygdala, the portion of the brain that is also responsible for the autonomic fight or flight response and, when damaged by repeated trauma, is also the biological trigger for post-traumatic stress reactions. Broca's Area, we also now know from the work of scholars like Noam Chomsky and Carl Sagan, is the seat of human language development—not vocabulary, but the structure of language itself. Thus, Edison was partly correct in his assessment of the importance of Broca's Area for cognition.

If, as Edison has written, memories, which are actually groupings of electrons he called "life clusters," have an existence independent of incarnate physicality, then they might survive after the death of the individual in a state not unlike a Nirvana and be able to be tapped. The science, therefore, would be to fabricate a machine delicate enough to identify the presence of those life clusters, then to discriminate among the groupings of those clusters to define an individuality. The next step would be to create a channel of communication wherein the living person sitting at the spirit phone can recognize the presence of those clusters through the photo cell meter. Would the life clusters understand English or another language? Edison wrote that he did not know this, but if the spirit phone worked, he would at least find out whether his theory of consciousness after death was verifiable.

Thus, the spirit phone became Edison's last great project, his magnum opus, to prove that even as he was entering his eighties and had become the founder of General Electric, he was still capable of coming up with an invention, a machine that could answer the age-old question about what happens when we die. Whether it succeeded and what became of it is a mystery. But the bigger mystery about the man who invented the twentieth century, who claimed to have shunned mystics and mediums as charlatans, who seemed to understand the consumer marketplace, and who translated everything he perceived into things palpably material, is why was it his quest to prove that his machine could contact the dead? This is the question we answer in this book. And so we begin with the forces that shaped young Tom Edison as he first learned to read, to reason, and to experiment with chemicals in his parents' basement.

CHAPTER 2

Young Tom Edison

"The secret of education is respecting the pupil."

—Ralph Waldo Emerson

Thomas Alva Edison was a man of his time who, through his inventive genius, superseded time by creating the technology of an entire century. In so doing, he bridged the age of the expansion of America in the 1840s to the age of communication and mass media.

EDISON'S GROWING-UP YEARS

Timeline
1840s: First telegraph comes to Milan, Ohio
February 11, 1847: Thomas Edison born
1850s: Edison fascinated by hot-air balloons
1857: Experiments with home chemistry set

EDISON THE PRODIGY

Thomas Edison, from the time he was a small child, evidenced the traits of a true American prodigy, ultimately curious and restless at the same time, hungry to explore the boundaries of his existence, and fascinated by the new age of technology blossoming around him. He would come to embody not just the age of invention and technology, an age of science-based industry, but the American dream itself, rising from humble beginnings in a canal town to a corporate entrepreneur whose son would become the governor of his adopted state, New Jersey. But none of this was apparent to Sam and Nancy Edison on the night their son Thomas was born.

Edison's birth was remarkably inauspicious. Nevertheless, the winter chill could not dampen the Edison family's excitement at the birth of a healthy baby boy on February 11, 1847, in their brick home on a bank overlooking the Huron River valley in the small northeast Ohio town of Milan.

Sam and Nancy Edison named their seventh child Thomas Alva, a lively infant with an adorable round face, dark eyes, and a shock of dark hair. But though he might have looked like many other babies, he was destined to be different. Even as a newborn, he was animated and seemed unusually fascinated by the world around him, by its colors and its sounds and even the touch of the fabrics that swaddled him. He was among the first generation of natural-born Americans in the Edison family.

Samuel Edison had left Canada for Ohio in 1837 to flee capture in a border rebellion. Nancy joined him in 1841. They settled in Milan, whose population had grown from 500 to 1,300 between 1849 and 1851, a growth attributed, in part, to the construction of the Milan Canal, which opened in 1839 and moved tons of grain, bushels of corn, and thousands of pounds of other produce and goods back east. Milan, on the Huron River, also became a center for shipbuilding, an important manufacturing center when the country was connected by a network of canals that required barges to move goods. However, once the railroads were built and prospered, more and more goods were shipped by rail, which was faster and cheaper. Like many river towns in the middle of the nineteenth century, Milan was bypassed by the railroads and lost commerce. The Milan canal's importance and revenues declined and businesses based upon the waterway transportation, building materials, and barge construction suffered. As a result, residents began to lose jobs and seek other places where they could find work.

That year, 1847, was not entirely good to the Edison family. It was a time when loss and bereavement were too-frequent parts of daily nineteenth-century life, when death often came at an early age. At a time of poor medical care, many childhood diseases were fatal. Infant mortality rates were high, and childbirth often resulted in mothers' deaths from infections and hemorrhage. In the same year Thomas Alva was born, three other Edison children died. As the population of Milan shrank, the Edison family joined the exodus, headed west, and settled in the larger community of Port Huron, Michigan. There, Sam Edison restarted the lumber and grain business that had at one time been successful in Milan. Because of the railroad's influence on growing towns and new cities, the demand for building materials was a growth business, and the Edisons made out well.

While Samuel tended to the business of building supplies, Nancy cared for her husband and children. This was a typical nineteenth-century Midwestern family, business oriented, religiously conservative, and upwardly mobile. She hoped that Thomas would receive some formal schooling in Port Huron, because he seemed very bright and inquisitive. The boy was no more than six or seven when his mother enrolled him in a local school run by a minister and his wife, who was the school's sole teacher. Typical of schools during that time, the classes focused on rote learning, religious studies, basic arithmetic, and reading. The lessons were mostly memorization and recitation of facts, things that did not excite young Thomas.

How qualified the woman was to instruct young children is a matter of opinion. There is no question that Thomas was intelligent and had a remarkable memory, but he was also restless and bored by the curriculum. However, his was behavior borne of curiosity. He was an inquisitive child, filled with ideas, reluctant to sit still in the classroom while his mind raced with questions and thoughts beyond his years. Nineteenth-century educational dogma believed children—like machines—should be disciplined and obedient to authority. They were not expected to question their teachers, only to answer their teachers' questions. Thomas did not fit comfortably into this unyielding structure of classroom conduct. He was simply too inquisitive, too anxious to find out what lay on the other side of the lessons he was being fed. He was too restless to sit still and simply absorb lessons.

One day, as the schoolroom tedium dragged on, Tom became fidgety and distracted. It was clear he was no longer listening, at which point the teacher took Tom's restive conduct personally. She blurted out that he was "addled," muddled or confused. Today a teacher might have referred to young Edison as unfocused or even suggested he had Attention Deficit Disorder. But back then, Tom's distractedness was considered a disciplinary problem. When Tom told his mother about his teacher's remark, she was furious, withdrew him from school, and, from that point on, homeschooled her son according to her own beliefs. His formal schooling had lasted only several months.

Mrs. Edison, who had once been a teacher, took over her son's education completely, introducing him to "natural philosophy," another term for science in the nineteenth century, and chemistry, which was his favorite subject. She also encouraged him to experiment on his own. Young Thomas was so excited by the sciences and his freedom to explore that he built his own laboratory in the basement of his parents' home. There he conducted his own experiments and read everything he could find about the sciences. Now freed from the constraints of the classroom, Tom was voracious in his reading and his experimentation with household chemicals. In fact, even before he was ten, he was performing small experiments to test the reactions of various chemicals he found around the house. But under his mother's tutelage and his own experimentation, he was actually educating himself, although mostly by trial and error, a process that would serve him well later in life.

Few children throughout history have emerged as geniuses at a young age, but it does occur from time to time. Wolfgang Mozart would of course fall into this category. In most cases, creativity is not recognized in a person until later years. In Edison's case, his mother appeared to be his beacon, showing the way on the first steps of his road to success. It was an exciting time to be on that road.

AGE OF SPIRITUALISM AND AGE OF SCIENCE AND INDUSTRIALISM INFLUENCES

Tom Edison grew up at a time when both the Industrial Age and Spiritualism had emerged as significant national social and intellectual developments that later influenced him. As we shall see, the two cultural waves would eventually crash into each other, merge, and result in a new age of science-based spiritualism in the late twentieth century, circa 1975. Certainly a young man as bright and sensitive as Edison was aware of both movements. America was in the midst of an industrial revolution, building railroads and factories, even as the Great Age of Spiritualism was sweeping the imagination of the country. As Edison studied his mother's lessons and kept notes on the experiments he was conducting, he had no way of knowing that someday he would become a world-famous icon blending those intellectual streams of thought—materialism, and spiritualism—into a single apparatus.

EDISON, THE HORATIO ALGER CHARACTER

The Edisons were a solid working-class merchant family. Edison's rise to fame as the quintessential industrialist, making him a very rich man, could be compared to Benjamin Franklin's rise from a printer's apprentice to a United States ambassador and elder states-man of his time. Edison's rise would turn out to become something of a poster-boy story, in the rags-to-riches tradition of Horatio Alger (1832–1899).

Horatio Alger, who wrote about how he achieved wealth and success in his autobiography, wrote popular nineteenth-century books for young boys that typically told of a youth who rose from poor or modest beginnings to become a "middle class and successful" adult by dint of his own motivation and hard work, sometimes with the help of an older, wealthy person. His works were an American version of Charles Dickens, his contemporary, who also wrote about upwardly mobile youths in novels such as *Great Expectations*. And Dickens's contemporary, former Royal Navy captain Sir Frederick Marriott, followed the same story line in novels like *Peter Simple* and *Mr. Midshipman Easy*: the rise of a young boy through his own pluck and intelligence to become a person of rank.

The term "Horatio Alger" became part of the American lexicon, characterizing someone who pulled himself or herself up by the bootstraps. We still hear this story today, told by political candidates who call themselves self-made individuals, because the idea is to live the American dream of self-fulfillment. And that characterized Thomas Edison. He would certainly have read the popular Horatio Alger books in the 1860s, the most well-known of which was *Ragged Dick*. Ironically, the real life Horatio Alger did not enjoy the success of his fictional character, not unusual for nine-teenth century-writers who were often poorly paid and unevenly treated. But it was the philosophy of the Horatio Alger mythos that influenced young men growing up in the latter nineteenth century.

Today, Tom Edison would have been called a child prodigy. However, in his own time, he was misunderstood by many as a willful, strange, and eccentric boy who followed his own path, was easily bored by traditional teaching, and was incautious when it came to pursuing his experiments. It must have been painful for him to be considered an outsider, especially by his peers. Fortu-nately, his strong will and interest in science allowed him to focus on the subject he loved the most.

INTELLECTUAL INFLUENCES ON EDISON

By the time Edison was reaching adolescence, science was also maturing. Nineteenth-century scientists could look back on hun-dreds of years of advances and a process of experimentation and reporting that ultimately became known as the scientific method, a process through which a theory could not only be proven eviden-tially, but demonstrated by repeatability. Edison pored over as many books as he could find, drawing on the works of such notables as Newton, Copernicus, Galileo, Faraday, Watt, Lavoisier, Whitney, Elias Howe, Ben Franklin, and Fulton. It was Francis Bacon whose confidence in the advancement of science encouraged the incep-tion of scientific societies.

Thomas Edison had plenty of material to read, study, and experiment with, plus his phenomenal memory allowed him to retain an enormous amount of knowledge. One of his favorite inventions, which he learned to use, was the telegraph, credited to Samuel F. B. Morse. That machine would play a large role in Edison's future. It not only provided him with his first paying job, it also started his thinking about communications technology, the field he would develop as an adult.

EDISON AND THE TELEGRAPH MACHINE

There is some confusion about when and where young Tom first saw a telegraph machine. Milan, Ohio, town records suggest it was in a telegraph office that was part of a local jewelry store. Tom was awed by this electrical machine with wires that clicked and clacked as it sent and received messages in a series of dots and dashes that represented words and sentences. Edison would learn the code, called Morse code, and would soon become an accomplished telegraph operator. While other children pressed their noses against glass jars of penny candy in general stores, Thomas was as happy to watch the Morse code transmitting over the telegraph wires connected to the machines and operated by young men controlling the apparatus. He was an "inquisitive dreamer," wrote author Carol Cramer in her biography of Edison.[3]

Samuel F. B. Morse (1791–1872) was an artist and inventor who received a patent for a system of code electrically transmitted as dots and dashes. Morse's dream of becoming an artist was rejected by the United States Congress, where he hoped to beautify the

---

3 Cramer, Carol, *Thomas Edison*, People Who Made History, (Farmington Hills, MI: Greenhaven Press), 2001.

capitol with his paintings. It was sometime in the 1830s when he learned about the telegraph. However, he knew little about electricity and so, although he wanted to build his own telegraph, he could not. Then he met a brilliant physicist named Joseph Henry, who agreed to help.

In 1837, Morse received a patent and $30,000 in funding to construct a telegraph line between Baltimore and Washington, a distance of approximately forty miles. He successfully completed that telegraph line in 1844. Morse's first message, "What hath God wrought?" was sent in what came to be called Morse code, the code that Edison would eventually master. To his shame, Morse never gave any credit to Joseph Henry. When the telegraph's use spread, it also came to Milan, Ohio, by the 1850s, which was when Edison first saw it.

The importance of the telegraph cannot be overemphasized. Until its first success, conversation or news between two parties or locations could only go as far as the two people talking to each other. Letters were the alternative, although they could sometimes take weeks to arrive, depending on where they were sent from and where they were going. But the telegraph allowed two or more parties to communicate over a span of many miles, distances previously unthinkable. The telegraph revolutionized the newspaper industry, government communications, the military, and, ultimately, personal communications via companies like Western Union. It wasn't long before telegraph lines were strung on poles across the country. The telegraph was literally the beginning of mass communication, bringing the country closer together. It was no small event. America was rapidly changing. Along with the railroad, the telegraph machine would play a large part in young Tom Edison's budding professional life.

EDISON, IRONWORKS, AND THE HOT-AIR BALLOON

When Tom was young, he was also fortunate enough to bear witness to the work of two other inventors. One was Zenas King, who worked tirelessly to develop an iron bridge. The other was Samuel Winchester, who, with his brother, was intrigued by hydrogen-inflated balloons and was eventually successful in supplanting hot-air balloons. Thus, Edison not only saw the beginnings of instant electronic communication, but also our first attempts to fly and to build iron-based structures. For a boy who was driven by curiosity and innovation, he was growing up in an age in which the former drove the latter.

Before he was a full-fledged adolescent, Edison was already hard at work learning Morse code and coming to grips with the promise of flight. These were new vistas for young Edison, the promise of instant communication across vast distances and the promise of flight, seeing the Earth from an entirely new perspective. By the time the nineteenth century had passed its halfway mark, the landscape had effectively changed, not just for Edison, who was already peeking over the horizon of a brave new world, but for society in general. For Edison, the conquest of the air and the promise of communication by wire, the traveling of signals close to the speed of light, meant that the world of canal barges and lumbering steam engines was on the cusp of change. And Edison would soon become the agent of that change.

CHAPTER 3

Trains and Telegraphs in Young Thomas Edison's America

"Stories of brave men, noble and grand, / Belong to the life of a railroad man."

—"Casey Jones," Nineteenth-century American folk song

Thomas Edison came of age, set up his first job, and made his first scientific discoveries at a time when the American continent was brought together first by canals, then by the railroad, and quickly thereafter by telegraphic communication.

From colonial times into the nineteenth century, America's modes of transportation consisted of either walking, horseback riding, or boating.

AMERICAN TRANSPORTATION MOVES FROM THE HORSE AND BUGGY TO CANAL BARGES

1820: The invention of Fulton's steamboat
1825: The Erie Canal completed
1859: Grand Trunk Railroad comes to Port Huron
1859: Edison's first business: butcher boy on the Grand Trunk Railroad
1860: *The Grand Trunk Herald,* Edison's publishing business, begins operation

1868: Edison moves to Boston as a telegraph operator

1869: Edison's electronic vote recorder invented

1860–1870: Railroads stretch from coast to coast

1869: Edison moves to New York

1870: Opens shops in New Jersey for telegraphic repair

1871: Edison marries Mary Stilwell

1873: Mary Edison born

1876: Thomas Edison, Jr., born

But then the country began to expand across the North American continent. By the time of Edison's birth, better roads and better means of travel were contributing to a stronger economy as goods and people were able to move more easily from one place to another. Better roads also meant that once-isolated trading posts could grow into new villages and then into towns. The next improvement in transportation centered on waterways, as well-engineered canals became transportation conduits for barges and other vessels. In 1820, the steamship SS *Robert Fulton* traveled from New York to Havana, the first steamship to make that voyage. Cargo boats, barges, and rafts traveled down the Mississippi to New Orleans. Along the various river towns that grew up, laborers would unload cargoes of grains and foodstuffs. New Orleans, already a diverse city by 1815, became a major transportation hub connecting the United States with Caribbean ports of call. However, while the journey along the Mississippi worked downstream, there was no navigable channel that moved north. That difficulty was eventually solved by steamboats that bucked the current and allowed for travel and commerce in both directions. By mid-century, the aptly named "paddle wheelers" could be seen stirring up and down the Mississippi. Some were described as luxury steamers where passengers not only dined and drank, but also danced and gambled into the early morning hours. Even today, redesigned paddle wheelers still ply the Mississippi, offering tourists riverboat cruises just as they did more than one hundred years ago.

Canal travel soon became part of early nineteenth-century life, connecting rivers and lakes. The Erie Canal opened in 1825, connecting the Great Lakes to New York's Hudson River and thence to the lower states and ultimately the Atlantic Ocean. With the completion of the Erie Canal, commerce blossomed even further.

However, the farther west one traveled, the worse roads were, with many being little more than mud-covered tracks marked by deep furrows. But that would quickly change when the Age of Railroads began. In 1830 there were thirteen miles of railroads in the country. By 1860, the year Edison turned thirteen, the number had increased to thirty thousand miles. By 1870 the railroad stretched from coast to coast. The upside was the growth of cities, jobs, the economy, and farming due to new machinery and equipment. The downside of the enormous growth in railroads was corruption and bribery, inasmuch as there was little government regulation in the years immediately following the Civil War. This lack of regulation left the industry wide open to a handful of clever and cunning financiers, such as Jay Gould, a notorious robber baron who seized control of the industry. It is likely that young Edison knew nothing about these machinations, despite his fascination with railroads—a fascination that would launch Tom Edison into his future.

In November 1859, the Grand Trunk Railroad of Canada set up a depot in Port Huron, Michigan. This gave the railroad an important connection, opening up Midwestern America to points north. It also became a hub for points west, particularly Chicago. In the ensuing decade, the railroad map expanded again, this time across the US, replacing canals as the preferred method of transporting goods. Railroads were routed to be closer to inland towns and cities, provided the basis for more expansive freight yards, and were a faster means of conveyance than mule- or horse-drawn barges.

Even during the height of the Civil War, the railroad industry continued to lay track from the East to the West and from the West to the East, hammering in the Golden Spike at Promontory Ridge in Utah in 1869, thereby connecting the entire continental United States by rail, a major industrial achievement. Thus, through a variety of connecting hubs, even the Edisons' town of Port Huron became part of the larger network of railroads. And because the train station in Port Huron was close enough to the Edison residence, young Thomas could walk there and stare longingly at the well-appointed depot building. His sights were far beyond Port Huron, and he longed to travel the rails to see new places and explore the possibilities of business enterprise. At that time, both the railroad and telegraph had extended in size and influence, and both fascinated Edison as they moved the country closer together, with both trains and telegraph lines crisscrossing the country in small towns and large. Young Edison began to experience wanderlust. The call of entrepreneurial enterprise was beckoning him.

EDISON'S FIRST JOB AS A RAILROAD "BUTCHER BOY"

By the age of twelve, Thomas decided to strike out on his own. Despite his mother's concerns, he obtained a coveted position as what was called a "butcher boy" on the Grand Trunk. Butcher boys, all young men, went up and down the railroad cars as the trains traveled from one town to the next, carrying trays from which they sold snacks, beverages, candy, newspapers, sandwiches, fruits, and small meals. Think of them as the first flight attendants selling snacks in airline cabins. They sold cigars, soap and towels for passengers to freshen up, and dime novels, which were the popular pastimes of the day and which romanticized the exploits of Wyatt Earp, Wild Bill Hickok, Buffalo Bill, Doc Holliday, and Bat Masterson. Some boys even carried a secret stash of magazines that could best be described as the nineteenth-century equivalent of today's pornography. From that and the snacks they sold, the butcher boys earned tips.

At first, Edison was simply one of the butcher boys. But that did not last long. The young entrepreneur not only worked the passenger cars, he became an employer of other youths, buying produce along the stops the Grand Trunk made and hiring young men to sell those pieces of fruit and other items to passengers, thus not only doing his own selling but profiting off the sales of others by creating his own downline sales force. This was a foreshadowing of how Edison would run his invention factories first in West Orange and then in Menlo Park, New Jersey.

The fares along the Grand Trunk were not expensive, even by today's standards, but the rides could be dangerous as the trains lurched and shook along tracks at speeds up to twenty miles per hour, all the while whistling, ringing their bells, and sending wood and coal smoke up and out of the train's chimney. There was always the danger of a derailment or a collision. Also, at first, brakes were manipulated by hand, which put safety solely in the hands of the engineer. Automatic brakes would not be invented and used until the 1870s.

The Grand Trunk railroad departed early in the morning for Detroit and returned late at night to Port Huron. Thus, Thomas had a long work day. However, there was a daily break of several hours during which he had plenty of time to read at the library of the Detroit Young Men's Society, a group he joined so he'd have access to their ample book collection. He recalled to friends years later that he had read every nonfiction book he could find. He was always a voracious reader, and was especially interested in books about science and history. His favorite scientist was Michael Faraday (1791–1867), both a physicist and chemist, and the inventor of the first electric generator in 1831, arguably the "greatest single electrical discovery in history," wrote Isaac Asimov in his *Bibliographical Encyclopedia of Science and Technology*[4]. When Tom earned enough money, he bought books written by Faraday.

FROM BUTCHER BOY TO NEWSPAPER PUBLISHER

Edison's strong work ethic showed itself in his many accomplishments, even at the age of twelve. One of his bright ideas—in the tradition of Benjamin Franklin in Philadelphia—was to publish a newspaper on the train and sell it to travelers on their bumpy three-hour journey. The enterprising youth pleaded with an editor at a Detroit newspaper for discarded type, ink, and other printing apparatus. Type was handset in those days, laid out in a matrix called a "chase," and locked into place on a press. With his pieces of type and a small, used printing press, Thomas Edison became a teenage reporter, editor, and publisher, likely the first one on a moving train in America. It was a great start to his lifelong entrepreneurial career. Where did he find space on a moving train for all his apparatus? Somehow he squeezed it into the crowded baggage car where he could set type, run the pages, and stack the papers before distributing them from passenger car to car.

---

4 (Garden City, NY: Doubleday), 1964, p. 361.

He called his newspaper the *Grand Trunk Herald*, and had a sub-scription price set at eight cents a month, thereby giving him a guar-antee of the number of issues to print and a stream of income to cover expenses and afford him some profit. In so doing he created a form of a subscription model that one can find in print and online magazines today. But a subscription business model entails a com-mitment to turn out the product people have paid for in advance, because a subscription payment is really just a loan from a customer that is paid back with the delivery of the item subscribed. It can be an onerous task, issue after issue, week after week. Ultimately, Edi-son found the work tiresome and began looking for a way out from under the obligation.

One of the reasons he'd kept the railroad job was so that he could give his mother some money every week. The Edison family was often hard-pressed economically, and Tom's income was a substantial help. In addition, offsetting the heavy schedule of writing and printing, the newspaper and butcher boy business afforded Edison the space, albeit tiny, in the baggage car. It was like having a private office, production facility, and, ultimately, a labora-tory all in one.

EDISON'S RAILROAD CAR CHEMISTRY LABORATORY

Besides printing his newspaper in the baggage car, which the train conductor allowed so long as Tom did not interfere with the management of the train, the young scientist also used the space to perform chemistry experiments. That's what got him into trou-ble, particularly when he decided to experiment with a new type of explosive he wanted to sell to the United States military—C_3H_5 $(NO_3)_3$, commonly known as nitroglycerin. One day, Edison mixed the key ingredients (nitric acid and sulfuric acid) into a beaker, let it cool to the appropriate temperature, and let the nitro compound rise to the top of the beaker. This he proudly showed to one of the military officers riding the train, who became so panicked when he realized it was nitroglycerin that he tossed it out of the window, where it exploded and almost derailed the train. In another experiment to create an explosive, something went terribly wrong and Edison's equipment accidentally burst into flames, setting the baggage car afire. That was the last straw. The angry conductor broke into the baggage car, extinguished the blaze, yanked Edison away from his chemicals by the ear (damaging it for the rest of his life), and evicted Edison at the next station. Off went young Thomas, all his chemicals and other materials strewn on the ground.

In one version of the event, Alexander Stevenson, the outraged conductor, was so incensed that young Tom Edison had let his curiosity overwhelm his caution that he lost his temper and punched him squarely in the ear, bursting his eardrum and causing the beginning of Tom's lifelong hearing loss. Then he grabbed the boy by the ears and tossed him from the train as it came to a stop. Another account, told by Edison himself, is that the angry trainman picked up him by his ears. At that point he "felt something snapping" inside his head. A third and more plausible explanation for Edison's deafness can be traced to an infection or a bout of scarlet fever he suffered as a child. Years later Edison's son, Charles, who became New Jersey governor, was also hearing impaired. Perhaps there was some genetic hearing impairment. Nonetheless, Thomas handled his hearing loss well, although it made him somewhat isolated from others. He did, however, learn to read lips, and regarded his partial deafness as a means of allowing him to concentrate on his work without outside interruption—he claimed he never heard a bird sing after he was twelve years old.

EDISON THE YOUNG HERO LEARNS TELEGRAPHY

Now fifteen, Tom was walking one day not far from a train depot. His eyes caught sight of a small boy, perhaps three years old, playing on the nearby train tracks, dangerous even on an empty track. Looking up, he saw a locomotive heading down the track toward the child with no time to stop. Thomas darted towards the track and scooped the little boy up in his arms to safety, just before the train rolled by. Tom had saved the child's life by putting his own in danger.

Moments later the child's father, stationmaster J. U. MacKenzie, came upon the scene. The father was beyond grateful to young Tom Edison for saving his son from certain death. MacKenzie wanted to reward Thomas, but he was not a man of means. However, Tom loved telegraphy and asked if he could learn telegraphic transmission and Morse-code skills from MacKenzie, a telegrapher. MacKenzie was only too happy to teach Edison. It marked a turning point in Edison's life, because it opened up a career path for him in communications. Tom had been fascinated not just by the telegraph itself, but by the promise of what he saw as instant communication from station to station and along a network. It demonstrated to Edison that the world of his parents—slowly moving barges down canals, letters that would take almost a month to arrive at their postal destinations, newspapers that often carried stories that had already gotten old—were all things of the past. The new world was the world of electronic messaging, dots and dashes that instantly connected people around the world by a network of wires. Whether it was by serendipity or meaningful synchronicity, this one event in young Edison's life opened up a path to mass communications, an industry which he would invent and reinvent in the twentieth century.

Once he had become proficient as a Morse-code operator, Tom left his job as an entrepreneurial newspaper publisher and butcher boy manager on the Grand Trunk. Armed with new and practical knowledge, he quickly found jobs as a telegraph operator in several Midwest towns and cities and in nearby Canada. When he applied himself, he was successful. But he was easily bored by the repetitiousness of the job, and as a consequence his career in telegraphy was an up-and-down affair. His skills were not the problem; it was his attitude and attention span. He could be careless, not always attentive, and sometimes given to daydreaming. Seemingly aloof, he did not get along well with his colleagues, and his mistakes got him fired on occasion. Some of his errors and what some called his antisocial behavior could be blamed on his deafness, but not all. His restlessness was a lifelong habit.

Another employment problem for Edison was the result of the Civil War's ending in 1865. When the troops returned to civilian life, many young men who had served in the army began applying for employment. Many of them had become accomplished telegraphers during the war, and as such were preferable to less experienced ones. Thus there were instances in which Edison was dismissed from telegraphy jobs through no fault of his own, but because he was replaced by more experienced operators trained by the military. However, he had persistence—a trait that would characterize him for the rest of his life—and when the job market leveled out after returning veterans had been absorbed by the industry, Tom found employment as a telegraph operator in Tennessee.

His "freelance" telegrapher career lasted roughly from 1863 until 1868, keeping him away from his family for most of that time. Yet he managed to return home once to visit his parents in Port Huron, at which time he was shaken to find them both in poor mental and physical health, living in near-shambles. It was a mess he did not expect. His beloved mother would linger at the edge of life until 1871, when she died, and Thomas attended her funeral. His father wasted little time remarrying. His new bride was a sixteen-year-old who had been a housekeeper for the Edisons at one time. She was more of a caregiver than a wife to the ailing Samuel Edison.

THE MOVE TO BOSTON

In 1868 Edison took a position as an operator at the main office of the Western Union Telegraph Company in Boston, a city rich in culture and educated people, where a number of inventors lived and worked. In the last quarter of the nineteenth century, Boston was what Silicon Valley and the Bay Area are today as centers of innovation. Edison at this point in his life was still deeply interested in electrical inventions, and in Boston he found several financial backers for his ideas to improve existing technology. In 1869 he resigned from Western Union and proclaimed himself a full-time inventor. That year, at age twenty-two, Edison received his first patent for an electric vote recorder, a precursor to today's electronic voting machines.

Although Edison's new invention was an ingenious device, he quickly discovered that the electronic voting machine would not be a success from a business standpoint. Even though it was simple to operate, the fact was that no congressmen were interested in a device that recorded a vote but did not allow for it to be changed or obfuscated. In other words, in 1869, politics were such that a vote you could not alter was a vote you did not want. Even today, in the wake of the 2000 election case of Bush v. Gore,[5] there is still an issue between hanging chads and an indelible electronic mark. Edison learned from his experience, though, and was reported to have

---

5 (531 US 98 [2000]).

said that it taught him an important lesson: never try to sell into a resistant or hostile market. If folks don't want it, they won't buy it, no matter how well it works. In the future, he vowed, he would first make certain there was a commercial market before he would invent a product or device.

While in Boston, Edison launched a new business, a commodity-quotation service with about two dozen subscribers. This enterprise was considerably more successful than his vote recorder, and it brought him some income and profit while offering his clients the most current price quotations for gold at a time when US currency was anchored to the gold standard.

Later in 1869, Edison moved to New York where he continued his work on inventions applicable to printing and automatic telegraphy. By 1870 he had opened two shops in Newark, New Jersey, specializing in telegraph repair, specifically printing and automatic telegraphy. There he could throw himself into his work, pursuing inventions that applied to telegraph operations. He also put Newark on the map, so to speak, as a center for telegraph manufacture and repair. This was important because it demonstrated that a new and profitable business could enhance the reputation of that location— look at Silicon Valley today. When integrated circuit chips were first conceived and inventors like Royce moved to the Bay Area, Silicon Valley as a high-tech hub was born.

EDISON STARTS A FAMILY

It also did not take long for Edison to notice an attractive young woman who worked in one of his shops. Mary Stilwell was just sixteen when she met Edison, eight years her senior. They took a liking to each other and began courting. However, Mary's father thought she was too young when Thomas proposed marriage. He wanted the couple to wait. And wait they did, until 1871. Tom and Mary were married on Christmas Day. Mary soon found out that the new groom proved to have been a more attentive suitor than husband. Once they were married, Tom was totally absorbed in his passion as an inventor. He ignored his young wife, who found herself alone for many hours at a time while Thomas worked.

In 1873 their daughter Marion was born, followed by Thomas, Jr. Their father dubbed them "Dot" and "Dash"—plays on the signals of Morse code.

Meanwhile, the inventor went on with his business. At some point he had about fifty part- and full-time employees. However, Edison did not like sharing credit because he believed the recognition should go only to him. In fact, his reputation did grow, and within the circles he moved in, many regarded Thomas Edison as a "genius" even though part of his genius was inspiring others and building on their ideas. Today there is still controversy about who really invented what, especially regarding the light bulb and the concept of moving pictures. The ideas might have come from others, as did the work of testing an invention, but the patents and claims of success all belonged to Edison.

It's clear from the work Edison did in his early years that the inventions he perfected, and the invention laboratories he opened first in New York and then in Newark, New Jersey, laid the foundation for the projects he would do when he reached his full potential. His efforts on existing technologies set the stage for him to advance those technologies beyond levels imagined by the original inventors, at a time when the currents of science and discovery were promising an entirely new century. Old beliefs of spiritualism were about to be transformed into a new culture in which science itself informed the world of spiritualism, and vice versa.

CHAPTER 4

Backgrounds to Edison from the Great Age of Spiritualism to the Age of Science and Industrialism

"There is no death! What seems so is transition; / This life of mortal breath, / Is but a suburb to the life elysian."

—Henry Wadsworth
Longfellow

In addition to the individuals who created their own times, think of those individuals like Gutenberg, Isaac Newton, Albert Einstein, and Sigmund Freud, the greatness of whose lives was also created by the times into which they were born. The needs of the environment propelled some of these individuals toward their greatest

TIMELINE

1847: Edison born into Great Age of Spiritualism
1848: Fox Sisters encounter the ghostly Mr. Splitfoot in upstate New York
1849: The reputation of Mr. Splitfoot's rapping spreads across the country
1849: Mr. Splitfoot identifies himself as a traveling salesman

murdered in the house the Fox family bought

1850: Editor and publisher Horace Greeley publicizes the reputation of the Fox Sisters

1853: President Franklin Pierce invites the Fox Sisters to the White House for a séance

1867: The death of Michael Faraday, discoverer of electromagnetism and direct current

discoveries and creations. Such was the case in the life of Thomas Alva Edison, born just one year before the beginning of one of the greatest intellectual movements in the United States and Europe, a movement that would shape Edison's thinking, as he both rejected it as an inventor and then tacitly embraced it in the last decade of his life. This movement was called the Great Age of Spiritualism, wherein the personalities that dominated it, some of whom wound up conducting spiritual encounters at the White House, and others who laid down the intellectual underpinnings for National Socialism in Europe, were also the builders of the cultural matrix into which Thomas Edison was born.

Ironically, the life of one of the most important scientists in the fields of electricity and chemistry, Michael Faraday, who discovered the principles of electromagnetic field induction, the basis for the theory and practice of direct current that Edison developed as an industry, overlapped the Great Age of Spiritualism by twenty years. Hence, even one year after Edison's birth in 1847, the Great Age of Spiritualism had begun, right during the greatest influence of Michael Faraday presaging the age of science and invention. These two streams of thought and discovery would ultimately influence Edison's work. And though both Faraday and Edison would dismiss spiritualism as fraudulent,[6] Edison would ultimately employ

---

6 Edison, Thomas, "Life After Death," *Diary and Sundry Observations of Thomas Alva Edison*, ed. Dagobert D. Runes, (New York: Philosophical Library, 1949).

the very people he disparaged as spiritualists to summon the spirits of the dead into his photon beam as proof of the concept of his spirit phone.

The Great Age of Spiritualism officially began on Friday evening March 31, 1848, in the modest wooden-frame home of the Fox family in a tiny upstate New York hamlet. Hydesville was just a group of homes and several stores and mills, typical of small farming villages that dotted the countryside.[7] The local newspapers and those who lived through that winter recalled it was one of the worst western New York State had ever experienced. "Biting winds, frigid temperatures, and snow battered the area."[8]

On December 11, 1847, just two weeks before Christmas, John and Margaret Fox and two of their daughters, ten-year-old Margaretta and seven-year-old Catherine (called Kate), both pretty girls with dark hair and expressive eyes, had moved into the unassuming rental house in Hydesville, New York. The Foxes had other older children who were living elsewhere with their own families. They were a close family, although Mrs. Fox was the friendlier, while Mr. Fox, a blacksmith, was largely impassive and laconic. Both the Foxes were religiously devout and prayed daily, relying on their faith to get them through the hard winter that froze western New York State to a standstill. However, despite their religious beliefs, they were unprepared for what would happen to their lives only a few months after they moved into the house.

From the previous resident, the house already had a reputation for unidentified and mysterious rapping noises, but the Foxes ignored

---

7 All references to the odyssey of the Fox sisters and to the Great Age of Spiritualism are described in Martin, Joel and William J. Birnes, *The Haunting of America*, (New York: Forge, 2009), 154–175.

8 Ibid.

the rumors. The previous tenants, the Michael Weekman family, claimed they were troubled by loud banging at the front door, but when they sought to answer it, no one was ever there. There was also the night when the Weekmans' eight-year-old daughter cried out that she "felt a cold and clammy hand pass over her face." The disturbances were sufficient to cause the Weekmans to move out after living in the apparently troubled house for less than two years.

IS THAT YOU, MR. SPLITFOOT?

The quiet of the Foxes' first few months in the house was broken during March 1848 when they heard noises they could not identify. Well, after dark one night, when the noises were especially loud, the Foxes searched the house by candlelight but could not find anything that might have caused the tumult. Mrs. Fox conceded she was frightened by the inexplicable rapping and mysterious sounds of footsteps. She drew her own conclusion about the noises: some "unhappy restless spirit" was haunting the house.

On the evening of March 31, 1848, the Foxes went to bed early, but it wasn't long before they were awakened by the puzzling noises. A search of the house again yielded no explanation. The children also heard the rapping and tried to imitate it by snapping their fingers. The children named the unseen entity Mr. Splitfoot because of their mother's fear that the disturbance was "the work of the devil."

Kate, the youngest daughter, commanded, "Mr. Splitfoot, do just as I do," clapping her hands. "The sounds instantly followed her with the same number of raps."[9] Now it was Margaretta's turn, and she said in jest, "Count one, two, three, four," hitting one hand against the other, and the raps answered as before.

---

9 Ibid.

Then Kate said excitedly, "Oh, Mother, I know what it is. Tomorrow is April Fools' Day and somebody is trying to fool us."[10]

Mrs. Fox agreed and tried to test the noise by asking it to rap out her children's ages successively. The rappings were correct. She asked it next if the answers were being given by a human being. There was no answer. Then she queried, "Is this a spirit? If you are, make two raps." The spirit complied with two loud sounds. Mrs. Fox asked the spirit if it had been injured. "If so make two raps." There were two raps in reply. "Were you injured in this house?" she inquired. Two raps answered her, indicating yes.[11]

Margaret Fox became alarmed, and called neighbors to her house. One was Mrs. Mary Redfield, who wasted no time telling her there were no spirits plaguing her and her family. She presumed the young Fox sisters were playing childish pranks. Mrs. Redfield conducted her own test. She asked the spirit questions. When she received correct answers, she became frightened and left.

Mrs. Fox sent for another neighbor, William Duesler. He questioned the spirit through raps and determined that a man, an itinerant peddler named Charles Rosna, was robbed of $500 and murdered in the house sometime in 1843 or 1844. If the spirit raps were accurate, the frightening revelation that the peddler's throat had been slashed and that his body had been buried in the cellar by a former tenant who had since moved away was unnerving to Mrs. Fox, who was struggling to reconcile her faith with what she was hearing from the rapping. Was someone still hiding in the cellar?

Several local men searched the cellar the next day, and by summer they'd found several human bones and some hair. The remains

---

10 Ibid.
11 Ibid.

seemed to have been buried in charcoal and quicklime to hasten the body's decomposition. There was no doubt that someone had been hidden there to cover up a death.

It wasn't long before news of the alleged spirit rappings caused a commotion in the neighborhood. Many people wanted to peek into the Fox home to see and hear the spirits for themselves. According to local news reports, as many as five hundred callers came in one day. Mr. Fox, a private person by nature, was not pleased with the commotion. "It caused a great deal of trouble and anxiety," he complained. "I am not a believer in haunted houses or supernatural appearances."[12] Still, the Foxes could not explain the noises, which grew worse and were soon transformed into moving objects: shaking beds, slamming doors, ghostly or etheric hands, the sound of someone being dragged down a flight of stairs, and "sounds of a struggle." The noises were originally heard at night, but eventually were heard both day and night.

There was no way news of the haunting spirits in Hydesville could be kept a secret, and it soon attracted both believers and skeptics who clamored to learn more about the Fox family and their supernatural experiences. Many were convinced that communication with the departed was akin to a spiritual telegraph.

In April, Mrs. Fox wrote a statement about the events that began in her home on March 31:

"I am not a believer in haunted houses or supernatural appearances. I am very sorry there has been so much excitement about it. It has been a great deal of trouble to us. I cannot account for the noises; all I know is that they have been heard repeatedly as I have stated."[13]

---

12 Ibid.
13 Ibid.

What had the Fox sisters done to ignite such a sensational response? They hadn't invented anything new. People had believed in—or at least longed for proof of—an afterlife since the time that humans first walked the earth. Ancient burial sites reveal the remains of the deceased in specific positions, suggesting they were facing in a particular direction, toward whatever gods they believed in, sometimes buried with earthly artifacts to pay homage to their gods. Every civilization has had its afterlife beliefs, though not all were the same. People, Edison and his rival Nikola Tesla included, have long sought confirmation of the spirit world's ability to communicate with the physical world. That is the basis of the belief in spiritualism.

The premise is not complicated. Our physical life is not long. However, if there is a continuation of life after death, then death is not termination. Rather, it is merely a transition to another dimension. And, if so, might some residue of the human spirit remain after death, some lingering sentience? This is what Edison believed. Finding incontrovertible evidence for life beyond death is difficult to do, but that is exactly what Edison sought. And that of course led him to the question of whether we can find evidence that the spirit world can communicate with those on earth. Spiritualists say yes. Atheists and agnostics insist the answer is no. For them, there is no afterlife. In between the two extremes are a variety of religious and spiritual beliefs that continue to be open to debate. Still, we have a right to ask where we came from, why we are here, and what happens to us after we die.

The Fox family had no idea that their experiences with invisible spirits, if that's what they were, had just begun, or that it would affect the whole nation. In fact, the Great Age of Spiritualism was just getting under way. It spread rapidly and by the 1850s writers such as Ralph Waldo Emerson, Herman Melville, Harriet Beecher Stowe, Elizabeth Barrett Browning, William Cullen Bryant, James Fenimore Cooper, and Henry David Thoreau were all influenced by beliefs in the spiritual nature of the afterlife. Others, including Mark Twain, did not hold to spiritualist beliefs. However, he did once admit that he had a dream premonition of his younger brother Henry Clemens's death later in a boat fire. The writer Nathaniel Hawthorne was also dismissive about spiritualism. Nonetheless, he reported seeing an apparition of an older gentleman, a minister, a number of times in the Boston Athenaeum. Hawthorne was surprised to learn that the elderly man had died before he had seen him.

When Louisa May Alcott's younger sister Beth fell ill with scarlet fever, she spent two years wasting away before she died. As Louisa and her mother gently tended to her at her bedside, Louisa wrote about her sister's final moments, when she and her mother saw a strange light appear on Beth's face. It ascended just as Beth stopped breathing. They asked Beth's doctor what that was, and he answered calmly it was most likely Beth's life leaving her body. The doctor added that he'd seen that same phenomenon many times in dying patients.

Even President Abraham Lincoln was an avid follower of the spiritualism craze in the 1860s. When his son Willie died from typhus brought on by the then-sewage-tainted water supply of the White House, first lady Mary Todd Lincoln was so distraught that she refused to leave her bedroom. Finally President Lincoln engaged the services of trance medium Nettie Colburn to summon Willie's spirit. Her séance, which apparently did summon up Willie Lincoln's spirit, so relieved the first lady that President Lincoln held multiple séances after that, often inviting members of his cabinet to attend.[14]

THE BEGINNING OF MESMERISM: TAPPING INTO THE UNSEEN

Franz Mesmer (1734–1815) was a German physician, mystic, and astrologer who believed that unseen forces made their way around the earth and influenced the lives of human beings. An early experimenter in the fields of electricity and electromagnetism, he believed he could cure diseases by manipulating these forces. In essence, he was a physician who was looking for unseen influences on human health.

Mesmer's first experiments used magnets that he passed over the bodies of patients and actually cured people in some instances. Later Mesmer found that he did not need magnets and the same healing results were accomplished by passing his hands over a patient. He termed this "animal magnetism."

Mesmer's patients experienced both successes and failures with his healings in Vienna. Eventually he moved to Paris, where he had a bit more success, but he came up against experts such as Benjamin Franklin, who insisted Mesmer's cures could be affected by the power of suggestion. Amidst controversy, Mesmer eventually retired to Switzerland and mesmerism became known as hypnotism, the term we use today.

We suggest that Mesmer was a possible influence on Edison because of his experiments in demonstrating that human physical

---

14 Colborn, Nettie, *Séances in Washington: Abraham Lincoln and Spiritualism during the Civil War,* rpt. (Toronto, Canada: Ancient Wisdom Publishing, 2011). See also Coburn Maynard, Nettie, *Was Abraham Lincoln a spiritualist?,* (Washington, DC: Library of Congress, 1891), and Martin, Joel and William J. Birnes, *The Haunting of the President,* (New York: Signet, 2003).

functions—he treated different forms of paralysis that did not have a physical cause—could be influenced by things we cannot see. It demonstrated to Edison later in the century that things unseen could influence things that were seen, just like electrons, which are unseen but perceived, were the conduit for information such as the dots and dashes of Morse code.

Mesmer's coming onto the scene in Europe decades before the beginning of the Great Age of Spiritualism also was a precursor to the belief that our lives could be the result of forces beyond our perception. In fact, Mesmer's experiments on his human subjects became a popular fare for other Mesmerists during the middle of the nineteenth century. Some of them even turned what Mesmer was developing as a medical approach into forms of entertainment and parlor games. Thus, the early nineteenth-century work of this German scientist helped set the stage for the coming of the Great Age of Spiritualism.

THE FOX SISTERS AND NOTORIETY

It had only been a few years since the Fox sisters had become a national sensation for the raps and cracking noises attributed to the unseen spirits that seemed to answer them. They demonstrated their purported psychic gifts to spectators and the curious. Of course, there were many skeptics, those who regarded the entire rapping episode as a hoax, a clever trick made by cracking their joints.

The Fox sisters nonetheless became quite famous, and even came to the attention of the noted newspaper editor Horace Greeley, who called upon them when they visited New York City. Greeley's story explains a good deal about spiritualism's value to the bereaved. Mr. and Mrs. Greeley had lost four of their five children from illnesses, and their grief was overwhelming. The most recent death was of their son Arthur, whom his mother Mary had believed to be a medium, and Greeley wanted badly to have spirit contact with him. He announced that the raps were genuine and not a thing the girls could have done. However, Greeley was not certain the sounds were from the spirit world. Nonetheless, Greeley publicly supported the Fox sisters, which not only promoted them but also lent credence to the entire spiritualist movement. Skeptics had a difficult time exposing the Fox sisters, which only strengthened their position as genuine.

The girls displayed their rapping talents before huge audiences, as well as at private sittings and séances for the wealthy and famous. Among the celebrities of their era included the writers James Fenimore Cooper, Willian Cullen Bryant, and Harriet Beecher Stowe. Their fame even took them to the White House, where they met President Franklin Pierce, the ancestor of former first lady Barbara Bush. For the president and the first lady, Jane, it was a fortuitous meeting. In January 1853, only two months before Franklin Pierce was to be sworn in as the fourteenth president of the United States, he and his wife were traveling by train with their only child, eleven-year-old Benjamin. Suddenly the train uncoupled and derailed. It tumbled, broke apart, and crashed down a rocky ledge. The Pierces suffered only minor injuries, but their young Benjamin was not so lucky, splitting his head open and dying before the eyes of his horrified parents. He was the only fatality of the train wreck. Mrs. Pierce never recovered from her son's sudden death.

To ease her grief, the frail and introverted Jane Pierce invited the Fox sisters to the White House in the hope she'd be able to make contact with Benjamin's spirit. This was the first time a psychic or medium was a guest at the White House. There are no records or notes of the séance for Mrs. Pierce, but rumors were that it was successful. It is also ironic that Barbara Bush's distant cousin by marriage, Jane Pierce, hosted a medium during President Ronald Reagan's term in office. First Lady Nancy Reagan requested it, and hosted astrologer Joan Quigley at the White House to determine the most propitious times for President Reagan to travel. When Vice President George H. W. Bush, perplexed by President Reagan's odd travel departure schedule, asked presidential aide Mike Deaver about it, he was informed that Nancy Reagan was receiving travel instructions from an astrologer. "My God," Vice President Bush was reported to have exclaimed. The irony is that his own distant cousin by marriage also hosted a medium who conducted séances at the White House.

Kate Fox also visited the White House when she attended a séance for Mary Todd Lincoln. Mrs. Lincoln was an avowed spiritualist and hoped to make contact with the spirits of her assassinated husband and their deceased sons. This tradition of paranormal and spiritualistic practices in the White House has continued up to the present. In fact, when former presidential candidate Hillary Clinton was first lady, she sought to channel the spirit of first lady Eleanor Roosevelt to help her navigate through the dark shoals of Bill Clinton's antagonists in what Hillary called the "vast right-wing conspiracy."[15] Why Eleanor Roosevelt? The answer comes right out of FDR's early days as president. On February 15, 1933, a very angry Giuseppe Zangara, claiming that he was fighting for the poor and the starving, and in a psychotic fury over the rich and entitled class, to which Roosevelt belonged, fired six rounds into Roo-

---

15 Martin, Joel and William J. Birnes, *The Haunting of the President*, (New York: Signet, 2003).

sevelt's open-topped limousine. He missed Roosevelt but hit five others including Chicago mayor Anton Cermak, who died from his wounds. And in 1934 a conspiracy among bankers, industrialists, and military veterans ostensibly led by Major General Smedley—who later that year disclosed that plot to a House committee on un-American activities—also set into motion a coup against FDR.[16] The anger directed at FDR was similar to that aimed at President Bill Clinton. It prompted First Lady Hillary Clinton to seek out Eleanor Roosevelt, as she had helped her husband deal with an angry constituency during his first term in the Oval Office. It was a fury that would be turned against presidential candidate Hillary Clinton herself over twenty years later.

---

16 Denton, Sally, *The Plots Against the President: FDR, A Nation in Crisis, and the Rise of the American Right*, (New York: Bloomsbury, 2012).

CHAPTER 5

The Second Industrial Revolution, the Science of Spiritualism, and Their Influence on Edison

It's easy to think of these two cultural trends as opposites. But to Thomas Edison, who was born into the pull between the world of industrialism and invention and the world of spiritualism, they were not. Although the first industrial revolution had begun at the end of the seventeenth century and continued into the eighteenth, the late nineteenth century also saw a wave of industrialism and invention, crisscrossing paths with spiritualism.

There is no record that Edison as a youth or young man had any interest in spiritualism. He was devoted to inventions, and most of his public statements dismissed as humbug spiritualist practices such as Ouija boards and séances. Yet Edison was aware of the strong public debate between the two sides of spiritualism, pro and con. Edison was at least curious about the nature of spiritualism, although he largely sided with debunkers and conservative or traditional scientists who scoffed at mediums and a spirit-filled afterlife.

55

At the time, Edison had working for him a young man named Bert Reese (1851–1926). Reese was an American-Polish medium who demonstrated clairvoyance to a high degree, and earned accolades from those who tested him. Edison watched and was impressed by Reese's ability. In fact, Edison even sat in at séances conducted by Reese. One detractor was the famed illusionist Harry Houdini, who branded Reese a fraud. But then again, Houdini never met a psychic or medium he thought was genuine. Throughout the Great Age of Spiritualism, accusations of trickery plagued mediums. In fact, many—if not most—mediums were ultimately uncovered as shams.

The one medium who remained incapable of being impeached as a fraud was Daniel Dunglas Home. Even Houdini's one-man war against psychics and mediums could not debunk D. D. Home, whose abilities included levitation. Could Home have actually risen from the ground and flown around a room or out of a window with no strings or other support? Houdini claimed he could duplicate Home's levitation, but he never did, which led some to believe that Home was the real thing. Over the course of his career, Home demonstrated a broad range of psychic abilities, especially his claims of prophecy and prognostication. On one occasion, in 1863, Home accurately predicted President Lincoln's assassination. It was a full two years before it occurred in 1865.

Another iconic figure of the Great Age of Spiritualism, whose influence spread across the centuries, becoming part of the Pan-Aryan underpinnings of National Socialism, and who remains important to this day, was Madame Helena Petrovna Blavatsky, a personage who dominated the late nineteenth century. There is no simple way of describing her. However, she is too important to omit from any discussion of the paranormal and mysticism in the nineteenth century, even if one only focuses upon her founding the Theosophical Society. Her reputation spread across the world, and was known to many celebrities and powerful people of her day. In fact, at one point she even crossed paths, however briefly, with Thomas Edison.

Born in 1831 in Russia, Blavatsky migrated to New York's Lower East Side in 1873. She was sometimes described as slovenly and overweight. She had fuzzy brown hair and a fiery temper, she dressed carelessly, and she chain-smoked. However, she had "magnetic eyes" that quickly captured attention. Two years after her arrival in the US, she founded the Theosophical Society, bringing Eastern philosophy and beliefs to the West. Her epic work was a two-volume massive compilation of her philosophy titled *Isis Unveiled,* published in 1877 and still in print.[17] (Isis was an Egyptian goddess and not the modern Islamic State.) Critics blasted it as "plagiarized poppycock" and a "heap of rubbish." In fairness, the twelve-hundred-page book did present a study of ancient religions amidst a good deal of confused writing mixed with her personal empowerment.

Blavatsky longed to bring celebrities and famous individuals into the Theosophical Society membership. She came close to convincing Thomas Edison to become a member, and he actually accepted a membership form to fill out, but then had second thoughts and pulled back, much to Madame Blavatsky's dismay. Edison did, however, keep a copy of Blavatsky's writings and, presumably, read them, although he did not write about them in any of his notes. Whether Blavastky was a genuine medium or a fraud, there is no denying her influence during the Great Age of Spiritualism.

---

17 Blavatsky, Helena, *Isis Unveiled*, rpt., (Pasadena, CA: The Theosophical Press, 1976).

Edison also read the works of the seventeenth-century mystic and clairvoyant Emanuel Swedenborg, with whom he was fascinated. Swedenborg's books, especially *Heaven and Hell* and *The World of Spirits*, originally published in 1758, were read by many nineteenth-century notables, including Ralph Waldo Emerson, John Greenleaf Whittier, Samuel Taylor Coleridge, Elizabeth Barrett Browning, and Henry Ward Beecher. Swedenborg's books were also in Edison's private library, but because Edison was sometimes self-contradictory when it came to his beliefs about things spiritual or metaphysical, all we have to go on today are his own words in his journals, which dismiss the world of spiritualism and its practitioners as frauds.

The Great Age of Spiritualism might have faded away by 1860 had it not been for the carnage of the Civil War. The bloodiest war ever fought on American soil claimed no fewer than six hundred thousand lives, perhaps more. The death toll breathed new life into spiritualism. The war had left thousands of widows, mothers without sons, and sisters without brothers. In other words, countless survivors grieving for loved ones killed in the war longed for contact with them. Mediums had a ready audience of the bereaved in the post–Civil War years.

In the immediate post–Civil War years, there were literally thousands of people who claimed psychic gifts. Most of them were women. Never before in American history had women been in such positions of power as they were when mediums commanded the séance table. A few of the women became famed mediums. Besides Madame Blavatsky and the Fox sisters, there were Florence Cook, Leonora Piper, Eusapia Palladino, Victoria Woodhull (the first woman to run for president of the United States), and Mary Baker Eddy, who became a famous healer and the founder of the Church of Christ, Scientist, although she despised spiritualism as the study of dark forces. For Mary Baker Eddy the meaning of "scientist" was one who studied verifiable knowledge, the original meaning of "scientia."

SCIENTISTS' QUANTITATIVE APPROACH TO SPIRITUALISM

By the 1880s, scientists, as we understand the term today, began to probe psychic phenomena seriously, forming the Society for Psychical Research (SPR) in England, and a few years later, organizing the American Society for Psychical Research (ASPR). The scientists made some startling discoveries suggesting psychic phenomena were genuine. Among the topics they experimented with or researched were apparitions and telepathy. Still, the skeptics and debunkers gave the SPR scientists no quarter. The critics of spiritualism and other forms of psychic phenomena were not about to yield. It did not matter that such eminent figures as the psychologist-philosopher William James were open to psychic events and manifestations, and willing to test them, which James did—and found evidence of genuine mediumship.

The Great Age of Spiritualism also produced some heated arguments between well-known individuals. One was the battle between the illusionist Harry Houdini, who despised spiritualism, and Sir Arthur Conan Doyle, author of *A Study in Scarlet* and other Sherlock Holmes stories, and an ardent believer in all manner of psychic phenomena. Another dispute grew up between two famed psychiatrists of the early twentieth century: Sigmund Freud, who professed no belief in the psychic, and his one-time associate Carl Jung, who found psychic events both genuine and useful. Both medical doctors, however, believed that the unseen world of the subconscious greatly affected the perceived world of the conscious.

Spiritualism was gradually fading by the beginning of the twentieth century, especially after one of the famous Fox sisters admitted publicly that their spirit raps were a hoax created by the girls snapping their joints, such as their large toe. But the spirits and mediums were once again saved by war. This time it was World War I, or the "Great War" as it was then called. The carnage produced by the terrible conflict included many thousands of victims, and thus numerous bereaved parents, siblings, and spouses. Once again, the bloodshed left many survivors, especially in Europe, seeking contact with the spirits of their deceased loved ones through mediums and séances.

One of the most famous and interesting books from the World War I era is Sir Oliver Lodge's *Raymond or Life and Death*.[18] Lodge's pursuit for survival beyond death became a serious effort after his youngest son, Raymond, was killed in 1915 in the war. After Raymond's death, Lodge visited several mediums. He had already seen Leonora Piper; now he visited Gladys Osborne Leonard, a well-known British medium, who told him she made contact with Raymond's spirit. Leonard described the existence of a group photograph Raymond had taken with other members of his military detachment, a photograph the Lodges had not seen and which Raymond, even in a letter sent them days before his death, did not reference. Yet, when the mother of one of Raymond's friends sent the Lodges the very photograph Leonard had described, Lodge came to believe that, indeed, Raymond's spirit had made contact. Lodge, a respected physicist, was also convinced that through the photograph he had made contact with Raymond's spirit.

---

18 Lodge, Sir Oliver, *Raymond or Life and Death*, (London: Methuen, 1916).

Parapsychologist Richard Broughton, in his 1991 *Parapsychology: The Controversial Science,*[19] explained that when the Society for Psychical Research was founded, applying scientific quantitative analysis to otherwise parapsychological events, it meant that a foundational change was in the offing for research into psychic phenomena. This was because psychical research could now become a disciplined science whose reliability was substantiated by experimental analysis.

The heyday of the Great Age of Spiritualism had come and would soon merge with the new science of the early twentieth century, creating a third culture that would influence Thomas Edison for the last decade of his life.

---

19 Broughton, Richard, *Parapsychology: The Controversial Science*, (New York: Ballantine Books, 1991).

CHAPTER 6

The Industrialization of the West and Its Influence on Edison

"It is not the length of life, but the depth of life."
—Ralph Waldo Emerson

When the first English-speaking settlers in America arrived here from Europe to settle the Jamestown colony in what would become Virginia in 1607, they came upon dense forests, lush greenery, streams, fields, and meadows. North America was home to a native population of some two hundred thousand indigenous peoples east of the Mississippi River, many of whom would be decimated by genocidal wars and the white man's diseases, notably fevers and

GROWTH OF INDUSTRIES

- Textiles
- Steam engine
- Manufacturing: iron into steel
- Telegraphy and communications
- Railroads
- Municipal gas supply
- Manufacture of Portland cement
- Glass manufacture

TIMELINE

Late 1700s to late 1800s: the
great migrations to the United
States
March 10, 1876: The invention of
Bell's telephone
1880s: Jacob Riis documents the
underclass in New York

smallpox. Native Americans, organized by tribes, made their livelihood mainly by hunting, fishing, and trading.

English-speaking settlements eventually were built on the unspoiled lands, often in the harshest of climates and most austere of environments. Notwithstanding the inevitable hostilities and clashes between native people and settlers, Europeans managed to build homes, plant crops, and raise families, introducing a rudimentary agricultural economy to the New World. Nearly everything had to be made by hand or with the simplest of crude tools. There was no technology to speak of. Throughout the 1700s and early 1800s, immigrants continued to migrate to the American colonies, which had become sovereign states confederated into a central government. Many of the settlers came from England, Scotland, Ireland, and later Germany.

In the eighteenth century, Great Britain was the world's leader in the development of new sources of energy for manufacturing textiles in factories. The movement was called the Industrial Revolution, and it followed the French Revolution in the 1790s. By the 1790s, small factories were already being built in a number of New England towns.

The phrase "Industrial Revolution" first appeared in print in 1799, written in a letter by a French diplomat who proclaimed that France had entered the push to industrialize its society. By the later decades of the nineteenth century, "Industrial Revolution" became a more widely used term.

In the 1830s, a decade before Edison's birth, several significant advances had already been accomplished in technology. Among them were advancements in textile manufacturing, iron making, and steam power.

Textiles: Formerly spun by hand, the entire operation of spinning cotton into thread and fabric had become mechanized via steam or water power, resulting in increased worker output. In applying power to what had been a hand-driven process, the textile industry became an early adopter of modern production techniques and moved the industry from private households to what would become modern factories. The Industrial Revolution saw a rise in the factory system, with fewer products made in homes and more manufactured in plants, mills, or workshops. Another advance was Eli Whitney's invention of the cotton gin, which in one day could remove more seeds from cotton bolls than a worker could do by hand in a month.

Iron-making: There was a significant change in the metal industries during the Industrial Revolution when wood and other bio-fuels were replaced by coal, which was more plentiful than wood. Utilizing coke for charcoal lowered fuel costs and made it possible to increase the size of blast furnaces.

Steam power: Steam engines became more efficient and used much less fuel. After 1800, the utilization of steam power grew greatly, and by the 1830s steam-powered boats were replacing sailboats. Steam would also replace water and wind power.

Another important industry that emerged during this time revolved around gas lighting, which allowed stores and factories to remain open after dark. This also resulted in some of the first municipal lighting utilities for cities, especially, New York, Chicago, and London.

Meanwhile, a new chemical process for the making of Portland cement was developed in the construction industry. The basic ingredient in the making of concrete, Portland cement was important for the advancement of the building trades by allowing a stronger material for the construction of taller buildings, tunnels, and sewer systems. Along with a new technique of producing glass, first advanced in Europe in the early nineteenth century, stronger concrete meant that cities could grow quickly.

Transportation methods also saw improvements during the Industrial Revolution, especially railroads to connect cities and carry freight more efficiently and quickly than barges or wagons. There was also an impact on agricultural production with the development of new farm machines and technologies.

Although the Industrial Revolution had begun in England, the birthplace of many modern technological innovations, it moved across the Channel to Europe, where new textile factories utilizing the power of steam-driven mechanics began springing up. Historians who study economics generally agree that the beginning of the Industrial Revolution was a critical step in human civilization, ranking with the domestication of animals and the development of communal agriculture. It spurred expansion, the creation of jobs, and population growth.

The Industrial Revolution marked a seminal pivot point in history because almost every aspect of daily life was influenced in some way. Notably, populations and incomes grew, resulting in an increased demand for food, goods, and services. Ultimately the standard of living itself was improved.

From 1840 to 1870, the First Industrial Revolution evolved into the Second Industrial Revolution. Technological and economic progress continued well into the next century, with the increasing adoption of steam-powered transportation (railways, boats, and ships), the large-scale manufacture of machine tools, and the escalated use of machinery in steam-powered factories and the development of assembly-line production.[20]

In America, the beginning of the Industrial Revolution had occurred in Boston, where Thomas Edison went to work for Western Union in 1868. Boston was a natural starting point for an inventor's career because it was a cultural and intellectual center dating back to colonial days. It attracted everyone from academic types to psychics and tinkerers, an early term that included many inventors. Edison enjoyed walking around the city meeting, watching, and talking to the tinkerers as they developed and fine-tuned their creations. He enjoyed the carnival of creativity that marked this early urban environment where he was free of mind. He could look at scientists in his own time for influence and motivation.

Thomas Edison had grown up during the so-called Second Industrial Revolution. It was an excellent time for an inventor to create and thrive. As Edison also found out, there was an increasing consumer demand for products, assuming he was able to invent them. It would be up to him—and other inventors—to determine what the best ideas would be. For not only did Edison need to create and finance a product, he had the responsibility to market the apparatus. As he became more experienced in analyzing the markets into which he was inventing products, he also came to realize that industrialization also required the migration

---

20 Atkeson, Andrew and Patrick J. Kehoe. "Modeling the Transition to a New Economy: Lessons from Two Technological Revolutions," *American Economic Review*, 97 (1) (March 2007), 64–88.

of people, many of them immigrants, into growing urban manufacturing centers where the jobs existed and companies needed cheap labor. This soon created an underclass of laborers who would someday require its own consumer market for Edison's inventions.

INDUSTRIALIZATION AND THE GROWTH OF THE UNDERCLASS

While industrialization brought about an increased volume and variety of manufactured goods and an improved standard of living for some, it also resulted in often grim conditions for the poor and lower working classes. Many of them were immigrants, crammed into horribly crowded urban ghettos, barely scraping by economically. One of the most fascinating—and chilling—documentations of the late nineteenth- and early twentieth-century slum life is a book of startling photographs and text, *How the Other Half Lives*, by Danish immigrant photojournalist Jacob Riis.[21] Riis was appalled by New York's Lower East Side when he arrived in America in the 1880s and documented it as a reporter for the *New York Evening Sun*, writing about the conditions of the poor and also working as a police reporter. He was an accomplished photographer and invented the modern conceit of reality photography as opposed to posed portraits. Riis was also one of the early press photographers to use the flash to light up shots in dim light as well as in interior photos. Jacob Riis Park in Rockaway, Queens County, is named after him because Riis, a social reformer, argued that the poor and underprivileged in New York lacked open spaces for recreation.

---

21 Riis, Jacob, *How the Other Half Lives*, (New York: Dover, 1971).

London also suffered from overcrowding and pollution as industrialization grew in the early and mid-nineteenth century. Child labor abuse was another problem worsened by industrialization, all of which was documented by Victorian novelist Charles Dickens in such works as *Oliver Twist* and *A Christmas Carol.* Edison was not a liberal, but as an inventor of products for the mass market, like the light bulb, was concerned about the state of the human condition. All of this had an impact on the way he viewed society and how he sold his inventions into it.

EDISON AS A MARKET-DRIVEN INVENTOR

For Edison, during this period, the importance of market-based research and development became paramount. After his experience with the failed vote recorder machine, he realized that he first needed to determine a potential market, then invent a machine or device aimed at a specific group of consumers. Inventions for the sake of inventing brought no profit, even though the joy of scientific experimentation did bring its own rewards. However, if one is to have an invention factory, which Edison later established at Menlo Park, New Jersey, one has to pay for it and one's patents have to have a value for licensing and manufacture. Other inventors quickly understood that. For example, Cleveland went from a city of 17,000 to 400,000 people, spurred by the manufacture of locomotives. The major American industrial city in the latter part of the nineteenth century was Pittsburgh, which had hundreds of factories, refineries, and foundries, and was the hub of steel manufacturing in the United States. The industry would sustain the city well into the twentieth century.

By 1903, the Ford Motor Company settled in Detroit and began mass producing low-priced cars, notably the ubiquitous Model

T. Chicago, the second largest city in the nation, had stockyards and steel manufacturing, as well as its shipping ports for the Great Lakes. The largest city was New York, a hub of immigration but also the center of the financial and banking industries. New York also became an important retail center, which by the turn of the twentieth century helped create an American consumer culture. Department store advertisements and catalogues dating from the late 1890s through the early 1900s show a remarkable increase in consumer products of all kinds. This would have been unthinkable only a few decades earlier. In this environment, Edison's electrical inventions thrived. Technology itself was propelled by new inventions, the growth of mechanized industry, and the speed at which new products could travel by rail, according to James Miller and John Thompson in *The National Geographic Almanac of American History*.[22]

The Edison model of identifying a consumer demographic market, defining a need for that market, and then inventing devices to sell into that market is a model still in place today. Think about the concept of office automation, touted by IBM in the late 1960s and into the early '70s. They identified the need for integrated office systems, devices that could talk to each other and share data and documents, then developed such devices and sold them into the market.

Think of social media in the twenty-first century. With just a small amount of training, people can communicate with one another and manipulate data on devices made specifically for that purpose without the overhead of technology getting in the way. This is all based on a model created in the early twentieth century by inventors and industrialists like Thomas Edison. Simply stated,

---

22 Miller, James and John Thompson, *The National Geographic Almanac of American History*, (Washington, DC: National Geographic, 2005).

Edison developed the technology to enable consumers to use his inventions without the need to understand or be trained in that technology.

Looking back from the vantage point of our technology-driven world, it's hard to imagine that until the middle of the nineteenth century, America was largely a country with an agricultural economy. It wasn't until then that there was a great migration from the farms to the cities. Jobs were plentiful, especially in the growing manufacturing economy. And with steady jobs came steady incomes. Steady incomes in turn increased the demand for consumer products, all of which was important for an inventor looking to sell his inventions.

For Thomas Edison, there was no shortage of ideas, only money. The need for financing his inventions was a key concern. Thus, he had to meet and interact with those affluent enough to invest in his projects, pitching them not only on the technology, but on the market for the technology. If an Edison invention was successful, the investors stood to make handsome profits. This came at a time when the US government had a hands-off policy on business, which created a class of people who became the masters of industry. Some of these people turned into robber barons, whose goal was to force out any competition, dominate the market, and manipulate the pricing structure of their products and services so as to create an economic oligarchy.

The formula for creating these vertical monopolies was not difficult. It was a matter of controlling raw materials and selling goods at lower prices until you forced out the competition and controlled a specific market, at which point you could raise your prices without fear of any competitors. One of the most powerful and fearsome robber barons was Jay Gould (1836–92), a railroad investor and financial speculator. There was also Cornelius Vanderbilt, who commandeered the railroad and shipping industries. Meanwhile, Andrew Carnegie took command of the steel industry and John D. Rockefeller created the oil trust.

It was also an era of several inventors rising above the rest. This list included Alexander Graham Bell, who devised the first telephone (1876); Guglielmo Marconi, who pioneered the wireless transmission of radio waves across the Atlantic in 1901, but who lost the US patent for it at the outset of World War II; the Wright Brothers, whose first airplane flew across the sands at Kitty Hawk, North Carolina in 1903; and Henry Ford, who created the assembly line for manufacturing automobiles in 1908.

Eventually, perhaps in opposition to the way the early robber barons ran their industries, laborers unhappy with low wages and poor working conditions organized into trade unions to improve their situation. A turning point occurred in 1911 when the Triangle Shirtwaist Factory fire in Greenwich Village killed almost 150 garment workers, mostly immigrant women between the ages of fourteen and twenty-three. Many of the workers were unable to escape the fire, because the owners had locked all the doors to prevent people from taking unauthorized breaks. That fire led to legislation to improve factory safety standards and helped spur the creation of the International Ladies' Garment Workers' Union. The labor movement did not go unnoticed by the monopolies and corporations, however. They hired strikebreakers, put together blacklists of union workers, and used armed guards to disband labor strikes, sometimes violently, and intimidate strikers. Labor-management tensions grew so bitter that between 1881 and 1905 about 37,000 strikes were called throughout the country, ultimately forcing the federal government to intervene. These are issues still being litigated today at both the state and federal levels.

The struggle between the owners of production and their workers created a societal tension that sometimes boiled over. Edison could not help but watch the social unrest unfold and wonder what role his inventions would play in the changing world. Otherwise, his thoughts and efforts were lost in his work.

AN UNDERCURRENT OF SPIRITUALISM

While the Industrial Revolution was marked by new industry and inventions, spiritualism was beginning to be examined scientifically. Many of its practitioners tried to employ technology in the belief that the living could communicate with the dead. This is when Edison, combining technological theory with spiritualism, realized he might be able to invent a device that could be used for communication with the departed. And that would be the spirit phone.

CHAPTER 7

The Third Culture of Spiritualism and Materialism and Its Influence on Edison

A NEW "THIRD CULTURE," A MERGING OF SPIRITUALITY AND SCIENCE

At the turn of the twentieth century, when Edison's productivity was near its height, there were two competing world views that defined the zeitgeist of the period. There was a world of the spirit that we could not see, of ghosts and entities that communicated with us. That belief in spiritualism was gradually overtaken by the age of science which, ironically, brought qualitative and quantitative analysis to spiritualism and thus influenced Edison to build his spirit phone. His goal was to determine whether the scientific theories of quantum matter and energy of life forms existed after death.

VRIL

The Great Age of Spiritualism that defined the philosophy of Helena Blavatsky and that enabled President Abraham Lincoln to assuage the first lady's grief over the loss of young Willie Lincoln did not completely fade away at the beginning of the twentieth century. Rather, after a brief period when it flourished, it then was transformed into an art form and a scientific principle. Indeed, Edward Bulwer-Lytton's novel, *Vril, the Power of the Coming Race*,[23] influenced Blavatsky and other late nineteenth-century thinkers to the point where there was a belief in the power of an unseen force, called the Vril, which could be tapped into by human beings. Is this like "Chi" as in "Tai Chi" or the "Force" of the *Star Wars* series?

The Coming Race was an early example of how belief in an unseen force, into which those initiated could tap, not only influenced readers, but influenced the founder of Theosophy. But it went further. Early in the twentieth century, a group of Russian women, Maria Orsic foremost among them, founded the Vril Society, which espoused a belief that not only was there an unseen force that humans could commandeer, but that the future of the human race depended upon the supremacy of the Aryan race. It was their belief that this race came to earth originally from the planet Aldebaran

---

23 Bulwer-Lytton, Edward, *Vril, the Power of the Coming Race*, (London: Blackwood and Sons, 1871).

and established itself in Thule, Greenland, and could tap into the Vril. Early Nazi propagandists seized upon this notion and replaced the Vril Society with the Thule Society and searched around the world for evidence of this extraterrestrial species. This search was the basis for the early Lucas/Spielberg *Indiana Jones* motion pictures, just like the Vril was partly the inspiration for the Force in George Lucas's *Star Wars*.

The point is that the Great Age of Spiritualism, having defined the latter half of the nineteenth century, spilled over artistically as well as politically into the early twentieth century. There it bumped into and, in effect, partly merged with the great scientific and artistic achievements of that time, the premise of which was that unseen forces govern the physics of reality. This is where it influenced Thomas Edison and his vision of a scientific/technological invention.

EINSTEIN, PLANCK, HEISENBERG, AND MEASURING THE UNSEEN

The great scientific thinkers of the early twentieth century were physicists Albert Einstein, Werner Heisenberg, and Max Planck, all of whom theorized about the nature of matter and energy, the relationship between the two, mysterious particles, how events taking place in the future influence events taking place in the past, and black holes. The theoretical physicists suggested that things that were not directly seen by the naked eye nevertheless existed and could be measured by some form of instrument, even if it had not been invented yet. This was important because these scientists argued that there was an unseen, but nevertheless palpable, reality governing the reality that we could perceive with our senses. In other words, the unseen governed the seen. Einstein went even further, theorizing, albeit after Edison's death, that not only were quanta of matter able to exert influence upon each other, even from a distance, but that one electron's spin could influence the spin of another electron, a type of cohesion of motion. In this way, electrons might have identical doppelgangers out there in the universe exerting influence upon one another. This theory was known as quantum entanglement or spooky action at a distance. Although Einstein was actually ahead of his time on this theory, it was an idea that partly influenced his spirit phone. It was also influenced by Einstein's 1916 General Theory of Relativity, which posited the existence of black holes whose densities were so great that when they collided, waves that curved both space and time, and even gravity itself, would be generated across galaxies. Such waves were actually measured in 2016, thus proving at least one aspect of Einstein's General Theory of Relativity.

FREUD AND JUNG: THE UNSEEN UNCONSCIOUS

Einstein, Heisenberg, and Planck were not alone in arguing that things unseen governed what could be seen. In the late nineteenth century, biologists, particularly microbiologists, discovered how unseen pathogens—unseen because they were microscopic—caused a series of infectious diseases, perhaps the most insidious of which was cholera. Also, at the turn of the century, medical doctor Sigmund Freud, in part fascinated by Mesmer's explorations into hypochondriacal symptoms, sought to explain how a part of the mind motivated his patients' otherwise inexplicable behaviors. What were the urges creeping into his patients' realities? What drove their fears, their lusts, their violence, their seemingly irrational actions? Freud came to attribute this to a part of a person's mind that was invisible to him, his subconscious or unconscious. We now take it as a fact that people have unconscious motivations that to some extent govern their conscious behavior. We even realize that actions that impacted us when we were children can still drive our behaviors today. In fact, the entire study of post-traumatic stress disorder involves how parts of the unconscious mind process the effects of trauma buried deep within a person's conscious. But in the early twentieth century, this was revolutionary thinking.

Similarly, Freud's student and colleague, Carl Gustav Jung, also believed in the existence of the unconscious as a motivator of human behavior. For Jung, however, there was not just an individual's subconscious, there was also a shared unconscious, one that was no less active in motivating behaviors and beliefs. This Jung referred to as a collective unconscious, a paradigm of symbology shared by different individuals across different cultures.

I'M OK, YOU'RE OK

Freud's and Jung's theories about the unseen element of an individual's consciousness and its effect on a person's behavior was borne out experimentally in the early 1950s at McGill University by neurosurgeon Dr. Wilder Penfield, whose research into the repositories of memory in the cerebral cortex demonstrated not only memories, but the immediate presence of events in the past. Think about Marcel Proust's description in his book *Swann's Way*[24] of the sensation of tasting a madeleine and having his mind flooded with memories of his childhood. The immediate olfactory sensation automatically stimulated his memories as if he were actually a child again. Penfield discovered much the same by stimulating a patient's cerebral cortex with a low voltage charge of electricity, so as to evoke

---

24 Proust, Marcel, *Swann's Way*, rpt. (New York: Penguin Classics, 2004).

the sensation that the person was actually there, reliving the event. Penfield's research was funded by the Rockefeller Foundation, those same folks who funded the Mercury Theater of the Air's 1938 radio broadcast of H.G. Wells's *War of the Worlds* as an experiment to see how Americans would react to a simulated extraterrestrial invasion of Earth. Penfield's experiments were also funded by the Central Intelligence Agency, which was seeking to use the unseen elements of the human mind to reveal the true identities of deep-cover Soviet spies who had infiltrated the United States right after the Korean War using false identities of captured American military personnel.[25] Thus, not only was the unseen aspect of human consciousness argued in medical theory, experiments to prove its existence were proven scientifically (and funded by the United States intelligence agencies as part of the MK-ULTRA project). Today, the pop psychology self-help books *The Games People Play*[26] and *I'm OK, You're OK*[27] were both based upon Dr. Wilder Penfield's work at McGill.

DECIPHERING THE UNSEEN IN ART AND LITERATURE

In the world of art and literature there was also a new trend gaining traction, a trend in which folks theorized that reality was something only perceived subjectively by individuals. Simply stated, different people perceived different things even when they were looking at the same thing. Perhaps one of the most exemplary pieces of literature involving the subjectivity of perception coming out of the turn of the century was Bram Stoker's 1897 novel *Dracula*. Films starring

---

25 Corso, Philip J. and William J. Birnes, *The Day after Roswell*, (New York: Pocket, 1997).

26 Berne, Eric, *The Games People Play*, (New York: Grove, 1964).

27 Harris, Thomas, *I'm OK, You're OK*, (New York: Harper & Row, 1963).

Christopher Lee or Frank Langella, or even the original 1931 version starring Bela Lugosi, don't depict that the original story was told in terms of different narratives, journals, and diaries, from different observers, each describing in his or her own terms what he or she saw. Thus, Jonathan Harker saw one thing, Mina Murray another, and Van Helsing still something else. The artistic conceit was that different people see things differently so that reality, far from being shared, was differently perceived. And more important, it was the unseen in Dracula that drove what was seen. For example, in the wild chase scene toward the climax of the novel, the pursuers must rely on Mina Harker's visions in her hypnotic state, within which she has an almost psychic connection to Count Dracula, to chase and find him. Only her mind was able to reach out, a different level of reality.

What was apparent in literature was also apparent in art and in music with the emergence of Cubism, the breaking down of an object and replacing it with a vision of that object in an abstract form, in some cases turning it inside out. The most celebrated of the Cubists, Pablo Picasso, articulated this vision, in which it wasn't what was objectively perceived by the artist that was represented on canvas. It was the artist's subjective vision, the unseen by others, which allowed the artist to abstract the subject. And in music, the Expressionist composers Alban Berg and Arnold Schoenberg did what Expressionist artists did with their canvases. Rather than recapture the way melody depicted reality, they sought to break it apart to represent what they felt rather than capture any external reality. In this way, expressionism and cubism both sought to expose the unseen, the subjectivity of the artist, to the audience.

Looking at a deep logic connecting what Heisenberg, Einstein, and Planck were formulating—the behavior of unseen quanta of matter affecting the universe; what medical doctors Freud and Jung were suggesting—unseen forces in the human mind controlling behavior, sometimes pathogenically; and what artists Stoker, Picasso, Schoenberg, and Berg were depicting, we can see there is an overriding convergence of thought. This was the belief system—that the unseen affects or controls that which is seen—to which Thomas Edison ascribed.

Comparing this to what spiritualists believed during the late-nineteenth century Great Age of Spiritualism, we see exactly the same thing: the unseen, things perceived spiritually or psychically, affects what is palpable reality. In this way, the cultural zeitgeists of the Great Age of Spiritualism and the new age of science and Expressionist art merged by the early twentieth century, exactly during one of Edison's most productive periods of invention. This convergence between the spiritual and the material of science redefined materialism, and must have been truly inspirational for someone like Edison, whose inventions relied on the results of unseen electrons traveling along circuits to their destinations. In fact, we know they were inspirational because, in 1920, the final decade of his life and in the wake of World War I, when all of Europe was marked by the heaviness of death, Edison embarked on his final invention, a communications device that could talk to the dead.

THE SPIRIT PHONE AND THE THIRD CULTURE

We suggest that the spirit phone represented an amalgamation of the work Edison had undertaken from his earliest days experimenting with telegraphy, through his creation of the recording and motion picture industries, and including his work on improving the telephone. It is almost as if the spirit phone is the objective correlative of the Third Culture. Edison had spent his entire life as the pioneer of communications technology, building upon his invention of the incandescent light bulb, essentially a filament tube, as a way to repurpose the conductivity of electrons. If you look at theories of subjectivity of representation, the presence of invisible forces governing reality, and the possibilities that electrons could maintain identical cohesion with one another across meaningful distances, and then combine that with the special theory of relativity that posited that matter could not be created or destroyed, only transformed from one state to another, we can see where perhaps the electrons that composed the consciousness of a human being did not perish with the death of the body, but lingered. And they could be measured.

This was not only the philosophical principle of the spirit phone, but the scientific principle as well. Even though Edison believed that most, if not all, trance channelers and mediums were hoaxers preying upon the bereaved, he thought perhaps there might be some science behind what they were preaching. If the Fox sisters and Nettie Colburn were onto some form of science, perhaps Edison could tap into it. And maybe, if he had a device that could register a clutch of cohesive electrons, he could prove the existence of a form of life after the death of the body. And, being a telegrapher himself, if he could develop a code, a form of yes/no responses, maybe, if that cohesion of electronic particles of life units still carried the sentience of the expired human being, then he could communicate with that mass of electrons that could not be destroyed, according to Einstein.

Though far-fetched, it would have been one of the most important breakthroughs in human scientific endeavor and typical of the man who invented the electric light and brought illumination to the world.

CHAPTER 8

How Edison's Inventions Led to the Spirit Phone

"Genius is one percent inspiration and ninety-nine percent perspiration."

—Thomas Alva Edison

Thomas Edison belonged to a select group of creatively gifted thinkers who, though they underperformed in school, changed the world. If you were to judge Thomas Edison by his accomplishments in school, you would be making a serious mistake. He belonged to a group of young people who either failed academically or had little formal education but moved on to great success. In this he was a lot

TIMELINE

1654: Blaise Pascal, French mathematician establishes the theory of probabilities

1867: Lord Kelvin's *Treatise on Natural Philosophy* is published

1900: William Henry Perkin and the foundation of the modern chemical industry

1900: Max Planck's quantum theory of matter and energy

1905: Albert Einstein's Theory of Special Relativity

September 1939: World War II begins

May 10, 1940: Winston Churchill becomes Prime Minister of the United Kingdom

EDISON INVENTIONS IN HIS EARLY PERIOD

1869: Electric vote recorder
1871: Universal stock printer
1874: Quadruplex telegraph

like Albert Einstein and Winston Churchill, both of whom had less than stellar academic performance records. We call these types of people prodigies. They have an innate ability, sometimes nurtured by their families, sometimes by others, not just to learn on their own, but to develop what they learn into important skill sets. What combination of innate intelligence and environmental influences produces such individuals remains as much a mystery as it does a controversy. And whatever the reasons might be, prodigies and geniuses fascinate the rest of us. A few worth noting would include:

- Blaise Pascal (1623–1662) was a French mathematician and physicist. Although physically ill, he was mentally very far advanced. Pascal was only nine years old when he discovered and then applied Euclid's first thirty-two theorems and organized them in their correct order. At sixteen he wrote a geometry book, and when he was nineteen he invented a calculating machine capable of addition and subtraction. This was in addition to his discoveries in physics, including the law, called Pascal's Law, that is the foundation of the hydraulic press, that showed that pressure on a liquid at rest in an enclosed space is transmitted without loss of energy in all directions. We would not have hydraulic braking systems in our cars were it not for Pascal's law.
- Sir William Henry Perkin (1838–1907), an English chemist, was only twenty-three when he became the most important

authority on dyes in the world. Perkin began working on dyes when he was eighteen. He'd always been interested in colors and the density registration of color so that textiles could be infused with color and would appear brighter.

- There was also William Thomson Kelvin (1824–1907), later honored as Lord Kelvin. He displayed the qualities of a prodigy even as an infant, and by the time he was eight years old, he was listening to his father's lectures on mathematics. He was eleven when he attended the University of Glasgow, and was only in his teens when he wrote his first paper on mathematics. He graduated second in his class. He was also an inventor who made improvements in the mariner's compass and tide predictors and many other formula-based theories. He was interred in Westminster Abbey near the body of Sir Isaac Newton.

Whether he knew it or not, Edison was a member of that exclusive group of geniuses whose names included Mozart and Einstein. Some attended universities; others barely had grade-school educations, and were largely self-taught. Few began life as wealthy young men. Edison fell into the category of men of modest means.

EDISON AS THE EMBODIMENT OF INTUITIVE THINKING

Thomas Edison is perhaps a prime example of a perfect storm of intuitive thinking, who combined all the essentials of creative development.

Author David G. Myers outlined "five components of creativity, the production of novel and valuable ideas."[28] The first is

---

28 Myers, David G., *Intuition*, (New Haven: Yale University Press, 2002).

"expertise." The second is "imaginative thinking skills." The third is a "venturesome personality." Myers offers as an example of this Edison's numerous attempts to find the correct material for his lightbulb filament. The fourth is "intrinsic motivation," and the fifth is a "creative environment." For Edison, all five elements of Myers' theory of creativity are quite accurate, and we can add to them a tireless work ethic and a Sherlock Holmes–like dedication to the principle of the process of elimination of what does not work. The process of elimination means that there is really no failure, only the homing in on what does work and the winnowing away of what does not. This dedication to find what does work characterized the work of Edison, who was absorbed in his labors for endless hours at a time, ignoring virtually everything else, including family, food, and sleep. His desk was as good a place for a power nap or a few hours' sleep as anyplace else.

Edison's intuition allowed him to identify the markets into which he could project his inventions, sensing what the public wanted and then developing a product to satisfy that need. This, too, is an element of creativity and intuition. At the same time, Edison's ability to understand market needs extended to how he worked with his laboratory associates and assistants. For example, as we will see, his work with his key assistant Charles Batchelor from the 1860s through the 1870s marked one of his most product-heavy creative periods.

Intuition is also defined as a flash of insight, a process of immediate understanding before the logic of reason sets in, a Malcolm Gladwellian moment of "blink."[29] Intuition has played a large part

---

29 Gladwell, Malcolm, *Blink: The Power of Thinking Without Thinking*, (New York: Back Bay Books, 2007).

in the lives and works of scientists, inventors, artists, composers, and other creative individuals. Often a highly gifted person has a difficult time explaining the genesis of a musical composition, a scientific discovery, or an invention. It is intuition that includes inspiration, instinct, premonition, or a sixth sense. Was it intuition that contributed to discoveries by inventors and scientists such as Thomas Edison?

Many songwriters have been asked where the idea originated for a certain song they composed. It is not unusual for the answer to be something along the lines of, "It just came to me." Johann Sebastian Bach reported that his musical inspirations appeared without any effort on his part. They just seemed to pop into his mind. If the same question had been put to Thomas Edison about the genesis of his ideas, what would his answer have been? He probably would have said that first you find a market that has a need, then you come up with a theory for a product that will satisfy that need. Then you reverse engineer the science so that, in effect, instead of proving something you don't know with science, you prove something you already believe to be true but have to establish its truth. This is how science really works and how Edison applied that process to the development of his products. But it started with an intuitive thought about market and need.

Myers cites a survey that found seventy-two of eighty-three Nobel laureates in science and medicine credited intuition.[30] One winner in medicine admitted that at times he felt as if he were being guided by an unseen hand. For his part, Pablo Picasso said that he was not alone in his creative moments because he believed a force was guiding him along the way, to help him find the art that was

---

30 Myers, David, G., *Intuition*, (New Haven: Yale University Press, 2002).

already buried in his mind. Like a master sculptor who uncovers the art already enclosed in the stone, Picasso peeled away the obvious to find the subjective truth.

Mental impressions, visions, or inspirations often come to talented or gifted people in dreams, providing ideas for inventions or even the subject matter for books. One successful author sleeps with a pad and pen near her bed so she can write down ideas for her novels that appear in her dreams. The nineteenth-century author Robert Louis Stevenson said that ideas for some of his books first came in dreams. More than one scientist in history has proclaimed that his dreams and visions led him to discoveries that he would not otherwise have made. One of the best examples was the great scientist and inventor Nikola Tesla, who was not afraid to announce that his visions often guided him to new scientific revelations. In fact, Tesla admitted, his visions were so vivid, in particular for the alternating current motor, that he merely copied what he saw in his mind as the blueprint for his inventions.

EDISON AS THE ANTI-SPIRITUALIST

While for most of his life Edison said little, if anything, about psychic or paranormal events, what he did say was typically against spiritualism. He specifically asserted that most of his successfully completed inventions were inspired by his previous failures, not only by flashes of inspiration. And although millions of Americans in the late nineteenth century professed a belief in communication with the departed, Edison considered the spiritual aspect of it "bunk," that is, until his time came to communicate with the departed—at which point he sought to find the science behind it.

Michael Faraday, the great British scientist who invented the electric motor, was particularly hostile to the belief in spiritualism that had swept across America and England in the nineteenth century. He was Edison's "favorite" scientist. As a young man, Edison hoped he could emulate Faraday in the number of accomplishments he'd achieve. Faraday also had little formal education; he'd left school at the age of thirteen. The controversial British naturalist Charles Darwin also eschewed psychic phenomena, although his later work did encompass the idea of some form of intelligent design.[31]

On the other hand, who knows what ideas percolated within Edison's mind as he grew older? His hearing impairment often closed him off to others, as did his introverted nature and near-obsessive work habits. If he did consider the supernatural, he might have said little about it to friends and colleagues.

SCIENTISTS AND THE PARANORMAL

There is a long list of famed scientists and other notables who were open to the paranormal throughout history. Among them were Isaac Newton, Emanuel Swedenborg, Benjamin Franklin, Alexander Graham Bell, Albert Einstein, William James, Carl Jung, Margaret Mead, Nikola Tesla, Joseph Banks Rhine, Louisa Rhine, Edgar Mitchell, and numerous others. Some of these names may surprise us. Few of us were taught that the great British scientist and mathematician Sir Isaac Newton studied both prophecy and astrology. Who knew that Benjamin Franklin had a curiosity about reincarnation or that he believed that people inhabited other worlds? How many learned that the American astronaut Edgar Mitchell had a deep interest in parapsychology, or about his

---

31 Darwin, Charles, *The Descent of Man, and Selection in Relation to Sex* (New York: D. Appleton, 1871).

absolute certainty that not only had extraterrestrial species visited Earth, but that our government knew about it and communicated with them?[32] Do many know that Einstein had books about mysticism and the paranormal in his personal library? Throughout the twentieth century, even in the face of professed skepticism, many thinkers tried to reconcile a belief in materialism, the argument that everything is made of matter, and spiritualism, that there is something nonmaterial in the universe. Edison sought to reconcile the two as well.

Despite his denials, Edison could not resist observing spiritualism practiced by at least one medium of his time, Bert Reese (1851–1926), who had earned a reputation as a gifted clairvoyant. He first demonstrated psychic abilities in public at the age of six, but because he was a child, many people in his native Poland feared the boy was either a "wizard" or was demonically possessed. By 1890, at the West Orange, New Jersey, facility, Reese had become one of Edison's assistants, and the great inventor actually held séances in private with Reese.

Reese's paranormal faculties once resulted in his arrest for disorderly conduct in the United States. To save himself from a jail sentence, Reese demonstrated his "powers" in court to the judge presiding over his case. Reese asked the judge to write three separate words or phrases on three separate pieces of paper, fold them, and press them in any order against Reese. Reese was correct all three times in stating the contents of the papers. This story, though debunkers have discredited it time and again, is probably one of the few moments when the existence of paranormal abilities was tried and proven to be true in court.

---

32 Private conversation with Dr. Edgar Mitchell.

Reese had his detractors. Most notable was the famed illusionist and psychic debunker Harry Houdini, who bragged he had exposed Reese as a fraud. That was not the opinion of Edison or many other observers, who felt that Houdini had overreached in his assertion. Houdini's claim that he'd debunked Reese is open to suspicion, however. It was well known that when Houdini could not disprove a psychic or medium, he simply lied and said he did, no matter how credible the medium, subject, or witnesses. His mission to debunk spiritualism and claims of the paranormal could be traced to Lady Doyle, the wife of writer Sir Arthur Conan Doyle. She had expressed her belief that Houdini's mother, Cecilia Weisz, a Hungarian Orthodox Jew, was asleep in the arms of Jesus, a statement which, for an Orthodox Jew, was considered anathema. This drove Houdini into a fury and made him dead set against anyone who claimed clairvoyant abilities. Coincidentally, both Reese and Houdini died in the same year, 1926.

EDISON'S EARLY INVENTIONS

Edison, as noted earlier, was more than an inventor or scientist. He also became a skilled marketing expert. After the failure of his first patent, the electric vote recorder, Edison vowed he would first determine the need for a consumer product or device before inventing and manufacturing it. In other words, an Edison invention had to have some commercial need or advantage, author Neil Baldwin explained in his *Edison: Inventing the Century*.[33] His second patent was for an improved stock ticker, one that saw his fortunes rise because it actually brought him some income. In fact, he received $40,000 from the telegraph companies, a huge sum in

---

33 Baldwin, Neil, *Edison: Inventing the Century* (New York: Hyperion, 1995).

the nineteenth century. The licensing of that patent launched him into the communications industry. Accordingly, many of his early patents dealt with the telegraph machine, which Edison continually upgraded and improved.

In his heart Edison was an inventor, and inventing was what he spent as much time doing as he could. With his crew of dozens of highly talented workmen in New Jersey, he constantly turned out new devices and found ways of applying new methods to existing technology. One Edison assistant who deserves special mention is Charles Batchelor, an Englishman who faithfully worked with Edison for twenty-five years. He was technically adept, self-effacing, and quiet. Others remained for various lengths of times, as Edison required their skills. With the success of Edison's reputation, his wealth also grew. In all, Edison was granted more than one thousand patents, an incredible number that stood as a record throughout the entire twentieth century. Yet it was his inventiveness in the communications industry that began the transformation of American society from a nineteenth-century small-town America into what became a nation united by mass media.

One can argue that it was because of Edison that, for the first time in the history of humankind, the "message outran the messenger," as Marshall McLuhan wrote a century later. Whether for military, news dispatches, business, or personal messages, the telegraph ushered in a whole new era of communications beyond anything roads or rivers could accomplish. In this world of technology Thomas Edison reigned as a crown prince. He was still in his twenties but he was the one in charge, not the others.

In 1874, at age twenty-seven, Edison finished his plans for the quadruplex circuitry, an ingenious device. According to author

Carol Cramer in her *Thomas Edison: People Who Made History*.[34] The quadruplex telegraph was not only faster than the original telegraph but, because it could send four messages out at the same time on a single line, it was substantially more efficient. The quadruplex was to telegraphy as today's email blast to a list is to a single email.

Edison was now becoming a man. Not yet thirty, he stood five feet nine inches tall and weighed between 130 and 135 pounds. Certainly no fashion plate, he often gave a rumpled appearance in his black suit and white shirt with or without a necktie. In 1876, he was twenty-nine and self-assured when he quit his Newark shop to build a laboratory in Menlo Park, New Jersey, with financial assistance from Western Union. Menlo Park, some twenty-five miles south of Newark, was America's "first industrial research laboratory."

---

34 Cramer, Carol, *Thomas Edison,* People Who Made History (Farmington Hills, MI: Greenhaven Press, 1966).

CHAPTER 9

The Wizard of Menlo Park and the Great Inventions

"Keep your shop and your shop will keep you."

—Benjamin Franklin

It is generally agreed that Edison's most productive period of creativity, his industry-creating inventions, took place at his Menlo Park research and development shop. There, the products he invented became the generators of entire industries. Also, devices like the motion-picture projector became the basic elements for what would become the spirit phone.

TIMELINE OF GREAT INVENTIONS

1745: The Leyden jar stores static electricity
1876: Electronic pen and mimeograph
1877: Carbon transmitter
1877: Phonograph
1879: Incandescent light bulb
1880: Municipal electric power distribution

In many ways, Thomas Edison was like Ben Franklin. Both were self-starters, both came from working/middle-class families, both were tradesmen—Franklin a printer's apprentice who became a publisher and Edison a telegrapher who became an industrialist—and both were the most important inventors of their respective generations who experimented with and discovered the power of electricity. Both believed in life after death and both believed that human beings were not the only intelligent life in the universe. But perhaps most importantly, both men were entrepreneurs who invented new industries. Franklin invented the lending library, the municipal fire department, and the United States Postal Service. Edison, in addition to the recording and motion-picture industries and the municipal power grid, was the founder of General Electric. But the strongest similarity between Franklin and Edison is that the two were individuals ahead of their own time. Franklin defined the Age of Reason in America while Edison defined the scientific and industrial revolution in America.

As he reached maturity, Edison knew he required larger quarters for the development of his inventions. He and his team, many of whom were consulting engineers, had spent about six years working from his shop in Newark, New Jersey. Among their inventions were improvements in telegraphy and, believe it or not, waxed paper. In 1875 the prolific Edison devised the electric pen, an instrument that allowed multiple copies of any document to be derived from a single stroke. It not only provided an efficient means for copying documents, but enabled workers to create multiple duplicates of documents. The electric pen became the ancestor of the mimeograph machine, a staple of offices large and small throughout the 1950s and '60s until replaced by the photocopying machine. When we think of the mimeograph machine today, and the way it empowered small printing companies and even political movements, we can see how that device alone helped to change the world by bringing information efficiently and inexpensively to the masses. Actually, Edison had substantial help from his loyal, long-time associate Charles Batchelor, but it was Edison who received credit for the device.

THE GOLD TICKER

One of Edison's first practical inventions for the financial services industry was an improved version of the stock price quote ticker, essentially a telegraph-like machine that sent important information about the value of stocks to subscribers. It was an early version of what became known as the wire service. Originally, however, the ticker was referred to as the gold ticker because it announced the fluctuating price of gold.

Edison's improved stock and gold ticker made its first appearance in 1869, but the story behind it is evidence of his opportunistic approach to mechanical and electrical engineering, which is how he originally climbed the ladder of success. As the story goes, Edison was walking through New York's financial district looking not just for job opportunities as a telegraph operator but to size up the market for new inventions. He had worked in the telegraph industry as a young man, but he had bounced from job to job, ending up in Boston, where, fired once again because he was tinkering with the very devices he was operating, he was broke. As a result, he traveled to New York with the intent of looking for a job as an operator. But for the excitement generated by New York's bubbling financial industry, the future might otherwise have seemed bleak indeed. Edison, however, was the eternal optimist.

In 1869, financial information was at a premium. The faster information could be generated, the more valuable the service became. At the center were price quotations concerning the precious metal investment commodity, gold, which were transmitted via a telegraph-like machine, a device that Edison had been trying to improve. Gold was the standard for currency then, and fluctuations in its price were an indicator of the economy and the value of the currency.

A fortunate moment in Edison's life occurred when he walked past Laws, the company that manufactured the Gold Ticker. This was a time when financier and investor Jay Gould was trying to corner the gold market while the American economy was plunging into crisis. The price of gold was fluctuating and the ticker at Laws, where Edison was sitting while waiting for a job interview, could not keep up with the price fluctuations. Finally, the quotation device stopped working entirely. It simply shut down, and the managers found themselves in a pickle. They could not report the price of gold at a critical time. The day was September 24, 1869, a day of financial crisis. While frustrated and harried maintenance workers tried to get the quotation ticker back on line, Edison, sitting right there, watched in fascination as attempt after attempt ended in failure. The thing just wouldn't work. Finally, Edison had watched enough. He stood up and asked for a crack at fixing it, telling the managers wringing their hands over the silent ticker machine that if he could take a look at it, he might find out what had gone wrong.

Frustrated and at their collective wits' ends, but amused at the offer from this young stranger, the men stepped aside and allowed Edison to inspect the device. What would be the harm in that? It had already shut down. But, sure enough, Edison's prior work with tinkering on telegraph machines paid off when he noticed that a contact had broken off and fallen into the depths of the machine. No contact, no current. No current, no transmission. Edison was able to fish out the tiny contact point and reattach it so as to repair the broken unit, and the machine started working again. What seemed like magic was merely a mechanical problem. The president of the company, who had invented the device, asked Edison whether he had the ability to improve the workings of the machine so that it would not break down again. "Of course," Edison assured him, and he was afforded his first job as an improver, not just an operator, of telegraph-like devices, at a good monthly salary of $300.

When Laws Gold Ticker Company was bought and merged with another company, the new owner asked Edison if he could make further improvements. Edison said that he could, and essentially invented what would become the stock ticker. He was offered not just a fee but a purchase price of $40,000 for the new device he had promised to invent. It was his first commission. For Edison, this marked the first time one of his brand-new inventions would fetch a purchase price, and it officially marked the beginning of the young man's career in practical inventions. He would go on to improve the telegraph machine and the telephone, but the improved gold ticker was the turning point in his life, affording him his first major sum of money.

THE AUTOMATIC PEN

The automatic pen worked by punching holes through the surface of a sheet of paper so as to turn it into a stencil. When coated with ink and then pressed onto another paper surface, the ink printed through, explained Neil Baldwin in *Edison: Inventing the Century*.[35] In this way, a single stencil could make fifty copies. This marked the beginning of relatively inexpensive document copying. With some smart advertising and publicity, the Edison electric pen sold well

---

35 Baldwin, Neil, *Edison: Inventing the Century* (New York: Hyperion, 1995).

when it went to market in 1876. The electric pen was the ancestor of the mimeograph machine, eventually marketed by the A. B. Dick Company. Those over a certain age will always remember the thin dark blue mimeograph sheets and the aroma of fresh ink and correction fluid from their public school years.

THE CARBON TRANSMITTER

In 1877, Edison tackled the carbon transmitter, a device that improved the original Bell telephone by turning the receiver into a microphone in which the carbon element converted the sound wave into an electrical signal. The carbon transmitter was an invention that not only improved telephone technology, but served as the basis for the creation of the carbon microphone, an essential component in Edison's later creations of the phonograph and talking motion pictures. But it was the improved telephone that proved to be a market success.

Although we credit Alexander Graham Bell with inventing the telephone, a marvel when it was first introduced by its inventor in 1876, it was far from a perfected device. The main problem with Bell's invention was that its sound quality was poor, requiring those using the device to shout into it, at least several times, to be heard. It also had a limited range, so that messages could only be transmitted a short distance because of resistance in the wiring that impeded the electron flow. Edison's goal was to improve the sound quality of Bell's invention, and at that he was successful because instead of using a magnet to generate the sound waves to the phone's speaker, he used carbon so as to vary the current. This removed the impediment problem, thus reducing the resistance, and not only allowed for a cleaner transmission of speech, but enabled the transmission to take place over a longer distance. The telephone proved to be one of the most remarkable inventions in the communications market, and launched the development of what today has become known as the "information appliance," the holy grail of the communications industry and the centerpiece of what will become, in less than a decade, human integration with a universe of silicon based AI-driven augmented reality.

In 1876, the United States Centennial Exhibition opened in Philadelphia, marking the nation's one hundredth birthday. There was a substantial display of Edison's inventions at the exhibition. Two of them won awards, one for telegraphy, another for the electric pen. As a matter of record the most popular demonstration was Bell's new invention—the telephone, the first successful instrument for the transmission of the human voice. The first telephone to go in the White House was during the presidency of Rutherford B. Hayes. It was so new that it wasn't unusual for callers to be answered by President Hayes himself. But it would be at the Menlo Park laboratory that Edison would perfect the sound quality of the telephone so that even the president could keep his voice down when he spoke into the device.

EDISON MOVES TO MENLO PARK

By 1876 Edison had a new address, about twelve miles south of Newark and some twenty-five miles from New York. He bought land and buildings in the tiny hamlet of Menlo Park, New Jersey, where it was to become the nation's first industrial research center. It cost Edison $5,200, not inexpensive in nineteenth-century dollars. His workers were his research team and his new enterprise would be an "invention factory," turning out hundreds of patented inventions. Edison combined electrical and chemical laboratories with an experimental machine shop and in so doing created the basis for today's R&D shops. There, Edison focused his laboratory personnel on looking at inventions he could sell into the communications market, since media had become his persistent fascination. Edison, like other inventors here and abroad, saw in the development of media a huge market for his inventions. What's so fascinating about Edison's focus upon the communications market is that today Edison's own company, General Electric, not only produces consumer appliances, but is one of the owners of media conglomerate NBC Universal. It is a fitting cap onto what Edison envisioned as the landscape of the modern world, a world united by electrical communications devices, as far back as 1876.

Edison biographer Paul Israel wrote that Edison's Menlo Park facility not only functioned as one of the most creative laboratories in the history of invention, but it also became the model for research and development facilities.[36] It was not just that it was an incubator of invention; it was actually transformative by corralling the creative processes of individuals with different technological specialties into a functioning type of assembly line. The Menlo Park facility also helped transform Edison from a single-minded nineteenth-century independent inventor to a twentieth-century industrial innovator and manager.

Edison received support for his laboratory from Western Union and later from his own Edison Electric Light Company to showcase to the world the value of creative invention for growing industries dependent upon innovation. Edison's laboratory eventually became the single most productive and largest privately owned research and development laboratory in the United States. It was certainly one of the most specialized facilities in the United States.

Structurally, the laboratory was a modest two-story frame building encompassed by a white picket fence. The building and grounds

---

36 Israel, Paul, *Edison, A Life of Invention* (New York: Wiley, 1998).

looked more like a pastoral setting than a lab whose contributions would literally change the world. Edison's loyal and closest associate, Charles Batchelor, also purchased a home for his wife and two young daughters in Menlo Park. The wives were not as happy in rural Menlo Park as their husbands because Newark was more urban, and New York City provided more amusement to the women and children, who were often left to their own devices by their husbands. Mary Edison described herself as isolated and neglected, and Rosanna Batchelor admitted she also endured loneliness and, out of fear, slept with a gun by her side.

THE PHONOGRAPH

In 1877, Edison devised the phonograph, the machine he considered his most significant invention alongside the motion-picture apparatus because it started one of the three most successful industries in American history. Ironically, today all three have merged into the wireless transmission of telephony, recorded music, and motion pictures with communications companies such as Edison's General Electric, Universal, and AT&T combining to deliver entertainment to consumers. Edison's two great inventions and his improvement on the telephone turned him into the inventor of the modern age of communications.

By 1878 Edison was a national figure, whose fame was akin to high-tech industry celebrities today like Steve Jobs, Bill Gates, Jeff Bezos, Elon Musk, or Mark Zuckerberg. Thanks specifically to his phonograph or what folks called his "talking machine," Edison astounded the marketplace with a machine that both recorded sound and reproduced it, including music and the human voice, and archived it on tin foil cylinders. Think of what a marvel that was in the 1880s as the phonograph became popular. The simple concept of recording and storing sound on a rotating cylinder so that it could be replayed over and over again was revolutionary. To some it seemed almost beyond belief that a machine was able to talk. It wasn't only the public that was stunned. So were many members in the scientific community, especially those who specialized in acoustics, a major area of research at the time.

Looking at these three pieces of apparatus, the telephone as improved, the phonograph, and the motion picture camera, it is easy to extrapolate why Edison, whose marketing genius foresaw the growth of the communications industry, devoted the final decade of his life to the spirit phone. From his perspective, it was simply another communications device. He probably saw that one of the last frontiers beyond communication with the living was communication with the departed.

The phonograph, too, was part of that media equation in Edison's mind because, in a way, and admittedly a stretch, it was a form of communication with the dead. Think about listening to artists who are no longer alive, like Cab Calloway or Billie Holiday, Frank Sinatra, Elvis Presley, Chuck Berry, Janis Joplin, or Prince, but who still communicate with us today on records or DVDs, digitally, and on film or video tape. Thus, Edison's belief in the market viability of recorded and archived sound was borne out.

No other inventor before Edison had attempted to build a machine to record sound. His was the first. And it evolved through stages, initially recording sound waves, transformed by a sensitive carbon receiver into vibrations sent to a stylus device that produced analog impressions on a piece of tin foil around a cylinder that rotated so as to memorialize a continuous stream of vibrations. He accomplished this by devising a way for his stylus to respond to sound waves by moving up and down in order to create a groove or furrow on the surface of the foil. When the foil was played back on the stylus, the grooves were retranslated into sound through a megaphone device that amplified it.

By the 1890s, the device had evolved again when phonograph cylinders were replaced by platter-like discs. In fact, in the 1950s, when discs were still in fashion, they were actually called platters by the disc jockeys on top 40 radio stations and by consumers who purchased them. Remember "The Platters"? Remember "My Prayer"? Remember the last dance at your senior prom? Today, original disc recordings, usually 33 rpm albums, are considered not just artifacts of an earlier age, but recordings for aficionados who appreciate the defects in the sound as opposed to the cleaned-up sound of digitized recordings. For the record, no pun intended, Edison's earliest words into the talking machine were "Mary had a little lamb." This he shouted into the receiver, and the machine reproduced it "perfectly," he was later quoted as saying.

It was Alexander Graham Bell and two colleagues who improved Edison's "tinfoil phonograph" to make it possible to reproduce sound from wax rather than tinfoil. Bell's invention was to impose or engrave sound waves into the wax with a needle or stylus which, just like the tinfoil device, moved back and forth in accordance with the vibrations coming through the receiver. The cylinders had a sound quality similar to the later discs, providing much the same audio fidelity, but discs had the advantage of being easier to mass produce by a stamping process, called "impressions," that remained in practice well into the 1960s. Discs are also easier to package in protective envelopes and easier to store either flat or upright, and are less expensive at retail.

On one occasion, Edison and associate Charles Batchelor were in Washington, DC, for a large conference when they were called to the White House by President Rutherford B. Hayes. What could have been the reason? Some matter of national concern? President Hayes was as fascinated by the phonograph as other millions of Americans, and so he requested a "private phonograph session." Mrs. Hayes joined them for the extraordinary late-night gathering.

Today, of course, the recording industry is a multi-billion dollar business, albeit one that is now not just a market of CDs and DVDs, but of MP3 transmission, including wireless transmission of sound, making it difficult to even imagine its humble gramophone beginnings. But, though digital today instead of analog waves pressed into a malleable surface, the basic premise still applies: translate the sound we hear into waves on a disk or ones and zeros passing through a circuit and play them back through a device that reads the code.

EDISON AS THE PUBLIC FIGURE AND BOSS

The public may have been intrigued and amazed by Edison's inventiveness. Many of those who personally knew Edison or had dealings with him often felt differently. Many saw a flawed person who was not particularly adept at social interactions. And there were reasons for that. For one, Edison could be aloof because his deafness seemed to close him off from others. He could also be self-centered, demanding, and a hard taskmaster. He was so focused on his dreams of inventions that he was not openly communicative with small talk or polite conversation. Today, we might call him a nerd and relegate him to a cubicle at a high-tech Silicon Valley development company instead of an electronics laboratory.

But Edison also understood the value of publicity and self-promotion, going so far as to put his own photograph on his phonograph cylinders. Thus it was rare for those who worked for him to receive credit for inventions they labored on. He could also be careless about his record-keeping. Employees might be paid on time— or not. Nikola Tesla was certainly a victim of Edison's reluctance to pay for work completed as promised. Still, Edison had little problem obtaining personnel because so many young men were eager to say they had worked for the famous Thomas Edison, whom they nicknamed "the Old Man."

For his part, Edison's hearing loss allowed him to maintain distance from anyone he wished to ignore. One group he chose to avoid was the hard of hearing. It wasn't unusual for him to receive mail from the deaf hoping he would invent a hearing aid. However, Edison did not think he could, and disregarded their pleas. He also spent so much time at work that he was rarely home. Politely, he was an inattentive husband. To sum up, Thomas Edison was a difficult and complicated man, a genius at one level, but a self-centered and demanding person in other ways. A British engineer once put it this way when describing Edison, "That young man has a vacuum where his conscience ought to be," noted Neil Baldwin in *Edison, Inventing the Century*.[37]

THE LIGHT BULB

Let's get this out of the way first. Edison did not invent the concept of the electric light bulb. He improved upon it. He invented what became known as the Edison tube, the staple for the transmission of current before the development of solid-state circuitry. And as Baldwin explained it,[38] Edison's improvements upon the electric light came at the end of a long and complex development history.

---

37 Baldwin, Neil, *Edison, Inventing the Century*, (New York, Hyperion, 1995).

38 Ibid.

Long before Edison or his associates gathered to invent products at Menlo Park, humankind sought a way to turn night into day by means of artificial light. Centuries went by with fire-lit torches, candles, oil, and gas lamps, all at various times serving to bring circles of light into the night. One problem was the danger of open fire, another was the fact that many light sources were not very successful in illuminating a large area. And all of them gave off smoke, sometimes black smoke that hung heavy in the air. Ever try to read by candlelight? Even oil and gas lamps are not very practical for lighting more than a small space after dark.

For many centuries, scientists were aware of static electricity, the electrical charge produced by friction. Perhaps some had already borne witness to so-called "earthquake lights," electrical flashes resulting from the static electricity generated by underground friction. The problem was how the electricity could be separately harnessed and channeled. The first significant efforts occurred by the mid-eighteenth century with the invention of the Leyden jar. Within the Leyden jar, static electricity was accumulated and then discharged in a single flash. But even the great American patriot and scientist Benjamin Franklin, who discovered that lightning had the same properties as static electricity when he flew a kite in a thunderstorm, could not find anything useful to humankind from electricity except to understand how it was transmitted in nature. Other scientists and inventors also tried to capture electricity for practical application, all with limited or no success even though they understood electricity was a form of energy. For example, British scientist Sir Humphry Davy employed a huge battery to display how electricity could be made to generate light. In its early days, many people found electricity little more than an amusing toy. But it was Edison who would change that toy into a utility.

Edison was already a celebrated personality when he began to tackle transforming the spark from a flash of electricity into a continuous current that burned a filament so that it glowed within a vacuum without catching fire. The light bulb as an idea had already been developed as early as 1802, but it was not a practical device. Edison sought to turn what was an oddity into a commodity. However, when he started work on it in summer of 1878, he was to discover that the process was more difficult and lengthier than he anticipated.

Although the fundamentals of light technology were generally understood by then, Edison and his team had a great deal to learn before they could turn the theory of transmitting electricity across a wire that would not burn up from the current into a practical device that could be marketed. That meant creating a heating element, a filament or wire that would not incinerate but would only glow, hence throw off light. The solution of the incineration problem was an understanding that in order to burn, a substance needed oxygen. Deprive an environment of oxygen and an element might glow from heat, but it wouldn't oxidize. That was the secret. To do this, Edison enclosed the entire circuit, the filament that would convey the current, within a vacuum protected by a glass enclosure, thus preventing oxidation.

Once Edison became excited about producing a practical incandescent light bulb, he worked tirelessly. His task was to find the perfect incandescent filament or wire that would heat up in an electrical appliance that could be vacuum enclosed within a glass bulb. Edison remained confident they could accomplish this practically and at a cost that would allow for mass marketing. His investors were not as sure as he was. One material he tried was platinum, but it proved too expensive. The team finally found the filament material that would work to their satisfaction: carbon. It was both plentiful and inexpensive and would last long enough during continuous use to make it an item consumers would buy. When Edison's light bulb was first demonstrated in 1879, it was immensely successful and widely praised. And the Edison bulb, the incandescent bulb, remains in use today, although governments are looking to phase it out because of the amount of energy it consumes.

THE INVENTION OF THE ELECTRIC GRID AND MUNICIPAL POWER DISTRIBUTION

In 1880 Edison determined how to distribute electric current across a system of cables.[39] By using underground cables, not unlike the way water and gas were distributed, electricity could be dispersed and metered. And because metering allowed for accounting for the use of the electricity, investors in power distribution could charge for it. Edison's first central power station was on Pearl Street in lower Manhattan. There, powerful dynamos or generators were built that would provide electricity to homes and offices. It was a long and time-consuming process to bring his goal to fruition. There were mishaps and delays, but finally the system worked. Edison declared he was satisfied, and today, "Edison" power companies like Con Edison in New York and Pacific Edison in California are the common names of the electric utilities.

At Menlo Park, under Edison's leadership and inventiveness, the Age of Electricity was born, and in decades ahead it would become accessible to people everywhere and transform not only the United States, but human civilization as well.

---

39 Adair, Gene, *Edison: Inventing the Electric Age* (Oxford: Oxford University Press, 1997).

CHAPTER 10

Edison, the Father of the Motion-Picture Industry: The First Step to the Spirit Phone

Edison's development of a projector to throw a narrow beam of light at a distance was the first step in assembling the components of what would become his spirit phone.

Thomas Edison's major inventions often built upon themselves to spawn new industries. Thus, along with the communications industry, Edison's most important and far-reaching innovation spawned the motion-picture industry which, almost from the first moments of silent movies, became the defining principle of modern society. It not only reflected society's values, it refracted them in such a way as to move society forward while defining the values of society itself. Compare D. W. Griffith's *Birth of a Nation* to Stanley Kramer's *Guess Who's Coming to Dinner* or the Mickey Rooney minstrel show dance number in blackface in Busby Berkeley's *Babes in Arms* to Gary Gray's *Straight Outta Compton* to illustrate this theory. And all this takes place within a single century.

One only has to look at motion pictures today to see how they have moved public consciousness, made us self-aware, and showcased values to which our society must aspire. Was this in Edison's mind when he fabricated the threading device that allowed a strip of images on sprockets to move along a spool in front of a powerful beam of light? Probably not, but inventions have a way of defining themselves and their own futures once they're launched, much like a virus seeking out new hosts and new ways to multiply and thrive. So it was with the motion-picture camera and projection apparatus, now squeezed into a digital format and pressed into a wireless telephone. This is also the result of Edison's improvements to the original device, so that the communications, recording, and motion-picture industries can be packaged in a small box, wrapped in a bow and sitting under a tree on Christmas morning. Quite an achievement. But there's much more because the basic premise of focusing a beam of light through a photograph onto a screen gave rise, in Edison's mind, to another possibility. And that was the spirit phone. Given the simplicity of the spirit-phone apparatus, it is not hard to see how Edison's development of the electronic motion-picture projector was its logical precursor.

THE PRE-HISTORY OF MOTION PICTURES

Projection devices were brought into existence as early as the seventeenth century by Athanasius Kircher, who called his device a magic lantern that projected images upon a surface. This was during the first Scientific Revolution. Thus, projection itself preceded by at least two hundred and forty years Edison's device. And as early as 1832, just sixty years before Edison began his process of motion-picture development, the first moving images were projected upon a screen. In 1882, Etienne Jules Marey created a camera that could take rapid photographs in sequence, which is what Eadweard Muybridge had shown only ten years earlier when he took photographs in a sequence of a galloping horse, to prove that its fore and hind hooves were off the ground at the same time. When viewed in rapid succession, Muybridge's photos actually showed the horse moving even though each frame was a still. That is how the eye, because of retinal memory or retention, played tricks on the brain, by making a series of rapidly moving still images look like an object in motion. You can try this yourself with a series of stick figures, gradually changing positions from page to page on a paper pad. It's called a flip book.

Although these inventions preceded Edison's camera, it would be Edison himself to conceive of a recording and playback device to manipulate and display visual images in the same way that his phonograph captured, recorded, and played back audio. But how to do it? The camera already existed that could capture still photos. And Eadweard Muybridge had demonstrated in 1872 just how a succession of images could be made to look like moving images. The question was how to assemble this into a single camera and single playback device.

By 1889, celluloid film had been created and was being used from time to time as a replacement for photographic plates. What Edison's associate William Kennedy Dickson figured out was that by punching holes along the edges of a roll of celluloid film, so that it would not interfere with the exposure window, he could align those holes with the prongs inside a camera so as to allow the film to track along a series of sprockets. Thus, he was able to create a mechanism of rapid exposures in sequence to give the appearance of motion. Now, having exposed the film, how to play it back?

EDISON'S KINETOSCOPE

Edison solved the projection problem by moving the film on a series of sprockets beneath a magnifying glass that was brightly illuminated from behind. A viewer could look through the magnifying glass at the film and see the succession of frames play before his or her eyes, thus, in effect, watching a primitive short motion picture. However, there was no projector, only a viewer. Edison called this device the *kinetoscope*, which he patented in 1893 and began showing off to the public. He constructed a studio for making his films in New Jersey and then planned to open a series of viewing parlors where customers could pay a fee and watch a short film loop. One of the first of these was called *The Sneeze*, starring one of the workers at Edison's laboratory. At first, Edison saw his kinetoscope as a novelty, a diversion, instead of a business, not realizing that he was on the cusp of creating an industry much like the recording industry. And as a result of Edison's initial short-sightedness, he did not register a foreign patent for his kinetoscope. Accordingly, unlike today's international intellectual property protocols, like the World Trade Organization, the Berne Convention, or the World International Property Organization, a failure to register a foreign patent meant that foreign developers could copy an invention without any liability for patent infringement and without the need to license the intellectual property. That's exactly what happened to the kinetoscope when foreign developers got their hands on the device.

IMPROVEMENTS TO THE KINETOSCOPE

In one adaptation and improvement, an English inventor, Robert Paul, who lived to see the expansion of the motion picture industry right up to World War II, developed a projection device that didn't just flip through the frames, but held the frame in place before the next one came up. Thus, his early films seemed to flicker on the screen, giving the modern slang name "flicks" to motion pictures. Across the Channel in France, the Lumière Brothers developed their own camera and projection device, which they, like Paul, were able to use to display images upon a screen. This meant that more than one viewer could watch a movie at the same time, and it opened up the possibility of motion-picture exhibition houses, or movie theaters.

THE VITASCOPE AND THE MOTION PICTURE INDUSTRY

Edison was, as we have seen, a consummate opportunist. He watched at first with fascination how his own invention, the kinetoscope, was overtaken by large screen projection units, which prompted him to develop his own projection apparatus. In the United States, Edison had left the movie-projector industry, but in 1895 the kinetoscope was replaced by the Vitascope, invented by Charles Jenkins and Thomas Armat. This popularized motion pictures and began to attract consumer viewing audiences. By the turn of the century, people were going to the first theaters. Other entrepreneurs became involved. Acting as distributors, they marketed their films to the independent exhibition houses. Thus, the entire film distribution industry began. It attracted entrepreneurs from other industries, including scrap metal buyer and later burlesque theater owner Louis B. Mayer, glove manufacturer Samuel Goldwyn, bookkeeper and owner of a nickelodeon business Carl Lammele, and streetcar conductor and later music printer Harry Cohn, all of whom became the first studio magnates, the patriarchs of Hollywood. With the expansion of theaters came the drive by the studio moguls to own the exhibition houses themselves, which they did, effectively creating vertical monopolies until the Supreme Court in *United States v. Paramount Pictures, Inc.*, (334 US 131 [1948]) effectively broke up the distribution monopolies, weakening the power of the studios and creating a power vacuum for talent agencies and their lawyers.

But as fascinating as this history of motion pictures is, more important for our purposes is its relevance to Edison's spirit phone. For those who are old enough to remember the audio-visual squads in elementary schools rolling motion picture or slide projectors into classrooms to show films, we might also remember that when the projection light was first turned on in a darkened classroom, some students created shadow figures against the light. They did this by interrupting the light beam with their hands, which projected shadows upon the screen. This was the basis of the projector in the spirit phone.

Edison, seeing how the screen displayed the interruption of the light beam, speculated that even dust motes, barely visible to the eye, could interfere with the beam of the light. If barely visible objects could interrupt the light beam, what about objects that were just as real but too small to be seen with the naked eye? Might these objects, if such existed, also be capable of interrupting the beam? And what if that were possible and the objects were invisible to the naked eye? Might there be another way to register the interruption of the beam? It was there that Edison hit upon the idea that instead of a screen to display the image interrupting the beam, an electronic registering device, displaying only photoelectric current, might act as a screen for something too small to be visible to the naked eye. And that device was a screen of sorts, only not the reflecting screen of the Lumière Brothers. It was a photoelectric cell that completed a circuit generated by the beam of light, actually a beam of photons hitting the electrically sensitive material. When the beam was interrupted, the photoelectric cell gave off its own signal, a weak current that would be displayed on a sensitive meter attached to the cell. And thus, the idea for a device to make palpable the unseen was born. It wouldn't come to any sort of proof of concept development for at least another twenty years.

CHAPTER 11

Edison vs. Tesla and the Battle of the Currents

One of the great events in my life was my first meeting with Edison. This wonderful man, who had received no scientific training, yet had accomplished so much, filled me with amazement. I felt that the time I had spent studying languages, literature and art was wasted; though later, of course, I learned this was not so.

—Nikola Tesla "Making Your Imagination Work for You"
The *American Magazine*, April, 1921

Creative genius and a keen sense of what products the market needed were not the only driving forces propelling Edison's career. Of importance as well, even if reduced only to pure business and psychological competition, was Edison's personal rivalry with Serbo-Croatian

TIMELINE

1856: Nikola Tesla born in Croatia

1863: Dane Tesla dies from a horse fall

1881: Tesla an inventor in Budapest

1884: Tesla moves to US

1884: Edison meets Tesla

scientist and inventor Nikola Tesla. Given their half-century of contentiousness and their battle over the business of municipal electrical-power distribution, it is not far-fetched to imagine that Edison's bold stroke to communicate with the dead was, in part, driven by his needing to one-up Tesla. At the time, Tesla was also developing a device to communicate with the dead; his was based on radio waves rather than a photon beam. Tesla believed that waves of light, sound, or even thought were eternal and traveled through the entire universe. It would be Edison, however, who constructed the first prototype of a device to communicate with the dead. It was his final conquest and ultimate competitive victory.

TESLA 101

From the time he was a child in Croatia, Nikola Tesla displayed a gift for being able to create images in his mind of devices that he wanted to design and then use those images as a building schematic or set of blueprints. Edison was different. With his concept for an electrical device at least lodged in the back of his mind,

Young Tom Edison, courtesy of US Department of the Interior, National Park Service, Thomas Edison National Historical Park.

Samuel Edison, courtesy of US Department of the Interior, National Park Service, Thomas Edison National Historical Park.

Helena Blavatsky, courtesy of Joel Martin.

Harry Houdini official postage stamp,
courtesy of United States Postal Service.

Bram Stoker, author of *Dracula,* courtesy
of BramStoker.org.

Gold price stock ticker, courtesy US Department of the Interior, National Park Service, Thomas Edison National Historical Park.

Telephone transmitter, courtesy of the Thomas A. Edison Papers at Rutgers University.

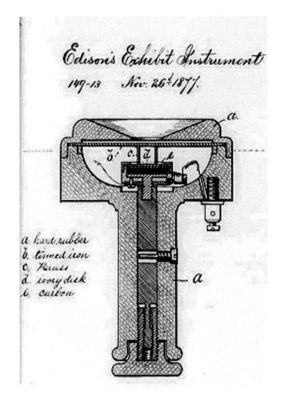

Edison's Exhibit Instrument
149-13   Nov. 26 1877.

a hard rubber
b, tinned iron
c, Brass
d, ivory disk
e, carbon

Menlo Park Laboratory courtesy of US Department of the Interior, National Park Service, Thomas Edison National Historical Park.

Edison at his tinfoil voice recorder (Edphono), courtesy of the Thomas A. Edison Papers at Rutgers University.

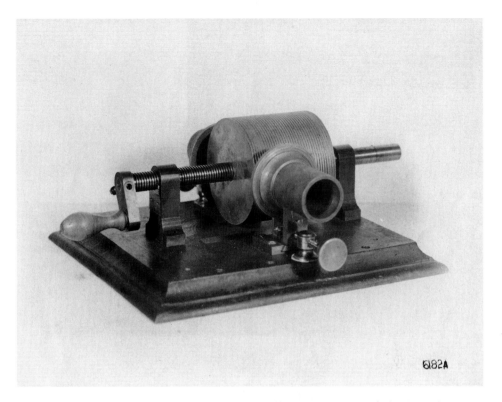

Wax Cylinder phonograph, courtesy of US Department of the Interior, National Park Service, Thomas Edison National Historical Park.

Edison at his projector, courtesy of US Department of the Interior, National Park Service, Thomas Edison National Historical Park.

Tesla portrait, courtesy of *UFO Magazine.*

Tesla in his laboratory (Tesla with machine), courtesy of *UFO Magazine.*

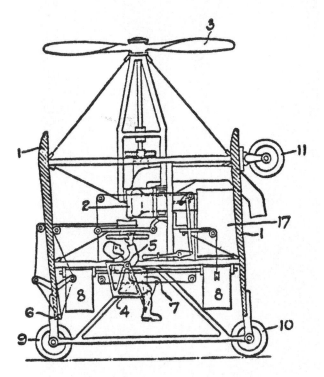

Tesla's Flying
Machine, courtesy of
UFO Magazine.

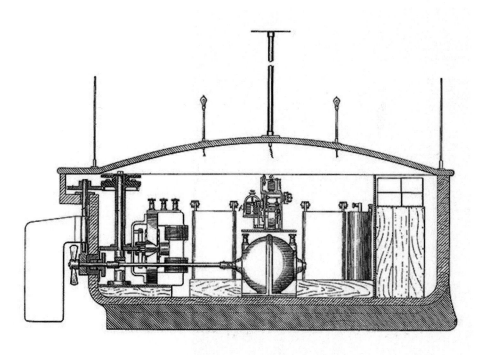

Tesla's Robot Boat, courtesy of *UFO Magazine*.

The Tesla transformer, courtesy of *UFO Magazine*.

Edsel Ford, Mickey Rooney, and Henry Ford after filming of *Young Tom Edison*. Photo courtesy the Monte Klaus collection.

Edison would let it percolate until he figured out the steps to build it. And this worked for him, until he crossed paths with Nikola Tesla.

As his primary industry expanded into commercial and industrial markets, Edison had to look at a serious professional challenge confronting him, one far more serious than he could have first imagined. This was the challenge posed by Nikola Tesla, whom he first met in 1884. Tesla's theory of alternating current generation was probably the most significant challenge to Edison's domination of the municipal power and light-generating industry. And for a small-town, homegrown, up-by-the-bootstraps inventor like Edison, whose power-generating stations in Manhattan were just getting under way from Pearl Street north, the appearance of Tesla in New York was a personal challenge as well as a business and professional one.

Edison was by all accounts a man who could hold grudges and translate a normal business rivalry into a serious life-and-death personal competition. Edison was jealous about the credit for his inventions, often taking the work of his lab assistants and patenting it under his own name. Thus, when Tesla popped up on the scene and refused to accept Edison's dominance of the invention industry, of the municipal power-generating industry, Edison not only took offense; he struck back, and hard. Perhaps it was Tesla's dashing appearance that first set Edison off. Or perhaps it was Tesla's unbridled confidence in his own engineering theories. Or perhaps it was Tesla's astounding ability to fix upon the design of an instrument in his mind and then use that design, as if it were projected on a screen, to recreate it on paper and then build it. Whatever it was, Edison and Tesla were almost genetically destined to become rivals in business and in life.

WHO WAS NIKOLA TESLA?

It is not a stretch to say that one of the most eccentric, fascinating, and complex individuals in the life and times of Thomas Edison was an engineer and scientific genius named Nikola Tesla, the developer of alternating current. Decades before a US Supreme Court decision yanked away the patent for the radio, over the objections of Marconi, who had sued the United States over patent infringement (*Marconi Wireless Tel. Co. v. United States*, 320 U.S. 1 [1943]), Tesla had competed against Edison in another area he had considered his own.

A Serbian born in Croatia in 1856 to Greek Orthodox parents, Tesla lived a life of frustration, always fighting for his patents, always dealing with skeptical financiers, a one-time millionaire who ultimately lost it all. Despite his genius, he never received the attention Edison knew in his lifetime. Ironically, in what became known as his battle of the currents with his former employer and mentor Edison, it was Tesla who came out on top, and whose vision of electrical distribution and supply is the one we use today. It was Tesla who most likely kept Edison pushing the envelope of new inventions and frontiers of knowledge, even as Edison reached old age. And it was Tesla's journals about his attempt to communicate with the dead via radio that challenged Edison to develop his spirit phone.

Tesla's father was a Greek Orthodox clergyman living in one of the most troubled and violent regions of Europe. The battle lines had been drawn almost a thousand years earlier between the Serbians and Kosovars, between Greek Orthodox and Roman Catholic, between Turks and Europeans, between Christianity and Islam. From this background, Nikola Tesla would become known also as the wizard of the twentieth century, an iconic figure depicted as the mad scientist in film (he was the cross between Flash Gordon's Dr. Zarkoff and Emperor Ming the Merciless), cartoons (his image appeared in the "Betty Boop" series), and comic books.[40]

The young Tesla, like Edison, was driven by curiosity more than he was inhibited by caution, a quality of his character that would mark him throughout his life. He was an explorer even as a school child, getting himself locked into inhospitable places, finding himself falling into vats of milk, and even experimenting with flight. He once used an umbrella as a parachute to see whether it could soften his fall when he jumped from a height. It didn't, and he was injured and confined to bed for almost two months.

TESLA'S EDUCATION

Tesla exercised his curiosity in books, principally books on mathematics and science, especially physics. When he attended the Croatian version of high school while living with his aunt, he saw his first modern machinery, specifically the steam engine. It was a wondrous sight for a young student completely enthralled by the mathematics of compression, propulsion, and the transmission of power. After suffering from a bout of cholera that laid him up for two years, he decided to become a professor of mathematics and traveled to the Polytechnic School in Austria. He also attended universities in Prague and Budapest, furthering his education in mathematics as well as in foreign languages.

Support from his parents eventually lagged and Tesla was forced to find work to support himself. He became an assistant engineer

---

40 Much of this section is based upon a Tesla autobiography from a publication of his lecture to the IEEE Society of New York in 1892 which can be found at www.gutenberg.org/files/13476/13476-h/13476-h-0.html

for the government telegraphy bureau, which enabled him to study the process of practical electrical engineering. However, his job, while it provided him with a salary, simply was not sufficient to sustain him or excite him. Tesla next turned his attention to what he thought would become a valuable endeavor: inventions and the ownership of patents. A patent for a device that people, specifically businesses, found valuable could eventually be used to get a steady income through licensing, something Edison had already figured out. But Tesla's first inventions were not for any designated business consumer market, and they brought him no fame or fortune. Less in desperation and more in hope, Tesla traveled to Paris and got a job as an electrical engineer building power-generating plants. There he also discovered that the center of scientific and electrical invention was not in Europe but in America, particularly in New York City under the auspices of Thomas Edison, whose inventions were changing the future right before his eyes.

By 1887, Tesla had established his own enterprise, called the Tesla Electric Company. (Today a company by that name, headed by Elon Musk, is making electric cars and storage batteries; developing a brain/computer interface called Neuralink for the upload/download/storage of human memory; essentially creating cybernetic organic hierarchical neural networks; and sending rockets into earth's orbit, then landing them on ocean-based platforms.)[41] With his new company, Tesla soon constructed and patented his first motor for generating multiphase alternating current, which was purchased by the Westinghouse Electric Company in July 1888. Thus, Tesla was on his way to recognition as an inventor of electronics in

---

41 Chapman, Glenn, "Musk Diving into Minds while Reaching for Mars," *Space Daily*, March 28, 2017.

America, especially when Westinghouse was able to sell his motors for a variety of industrial purposes. Tesla continued his research into alternating current motors even after he no longer worked for the Westinghouse Company, delving into other uses for high-frequency alternating-current motors. In an address to the IEEE in 1892, Tesla said, "Of the various branches of electrical investigation, perhaps the most interesting and immediately the most promising is that dealing with alternating currents. The progress in this branch of applied science has been so great in recent years that it justifies the most sanguine hopes. Hardly have we become familiar with one fact, when novel experiences are met with and new avenues of research are opened. Even at this hour possibilities not dreamed of before are, by the use of these currents, partly realized. As in nature all is ebb and tide, all is wave motion, so it seems that; in all branches of industry alternating currents—electric wave motion—will have the sway."[42] One can only imagine Thomas Edison's reaction to this prediction. It bluntly challenged Edison's own investment of his skills and business in direct current electrical motors, which he hoped would become the backbone of suppliers for what would become the twentieth-century municipal power grid.

In many instances, Tesla's abilities and inventions were nothing short of extraordinary. His mind was given to inventions even when he was a boy. He was only nine years old when he built what he called a "sixteen-bug power motor" by fastening June bugs to a thin wooden wheel so as to propel the wheel into a primitive drive mechanism. However, superseding all of his abilities, his unique mental power was his paranormal visions. Throughout his life,

---

42 Tesla speech to the IEEE, 1892.

"ideas flashed into his mind as working units, complete to the final details of component design and size."[43]

EDISON'S AND TESLA'S DIFFERENT EDUCATIONAL BACKGROUNDS

Unlike Edison, who had little formal education outside of homeschooling by his mother, Tesla had substantial schooling in mathematics, physics, and mechanics, and he later studied philosophy at the University of Prague, in the capital city of the Czech Republic. In 1881, Tesla began his career as an inventor in Budapest, Hungary, where he mainly worked in engineering for the Budapest Telephone Exchange. There he devised an apparatus called the "telephone repeater," an invention akin to today's loudspeaker. But Tesla, who had followed Edison's work, had a desire to work for him and share with him his idea for alternating current. In 1884, he moved to the United States and settled in New York City, eventually becoming a naturalized American citizen.

Tesla later credited his mother for encouraging his inclination toward inventions. She'd invented different types of kitchen utensils and was innovative in coming up with solutions to household problems. What's more, her ancestors were also inventors.

In 1863, the Tesla family was beset by tragedy when Nikola's older brother, Dane, was thrown from a horse and later died of his injuries. Nikola was only seven at the time. The Teslas were inconsolable in their grief and languished over the loss of their eldest son.[44] From then on, Nikola received little praise or attention from his parents. As a result, Tesla revealed that he had little confidence

---

43 As explained by Melton, Gordon, ed., *The Encyclopedia of Occultism and Parapsychology* 5th edition, (Farmington Hills, MI: Thomas Gale, 2000).

in himself growing up, something he overcompensated for later in life. Once, as a lonely child, he ran away from home.

Sometimes Tesla awoke in the middle of the night after nightmares of Dane's death, to which he said he'd borne witness with his own eyes, and of the funeral. No matter how hard he tried to avoid the visions, he was unable to erase them. "I was oppressed by thoughts of pain in life and death and religious fear… swayed by superstitious beliefs, and lived in constant dread of the spirit of evil, of ghosts and other unholy monsters of the dark," he wrote in his journal.[44]

TESLA AND THE PARANORMAL

There has long been considerable controversy about Tesla's visions and other types of paranormal events. Skeptics and traditional scientists have worked hard to find earthly rather than supernatural explanations for Tesla's visions and out-of-body experiences, often to no avail. It appears that no criticism or dismissal of Tesla's psychic encounters could alter his paranormal "gifts." On the other hand, Tesla always had his followers, those who were unshakable in their faith that he was telling the truth about his preternatural or metaphysical involvement. Ironically, Tesla's harshest critic about the occult and related topics was Tesla himself. He did not want to admit to his visions or other paranormal encounters for fear that he'd be ridiculed or associated with fraudulent mediums, of which there were many. We should remember that throughout the centuries, many scientists were interested in or at least curious about psychic phenomena and metaphysics. Many scientists studied, read,

---

44 Ibid.

and contemplated the occult, often without reaching any conclusions about its origins or nature.

TESLA ARRIVES IN NEW YORK

Tesla arrived in New York by ship in 1884. He was twenty-eight years old, tall and slender, with a lean, youthful face and a mustache. Tesla first met Edison's closest associate, Charles Batchelor, who introduced him to Edison via a letter. Meeting with the "Wizard of Menlo Park" was a thrill for Tesla, who had been anxious to meet him ever since he was a teenager. The result was that for nearly a year Nikola Tesla worked for Edison, who was initially impressed by Tesla's intelligence, skill, and ability to solve engineering problems.

The two men could not have had more different appearances. Edison, about five feet nine inches tall, did not appear to pay much attention to style. His dark suit was typically rumpled; his white hair, as he grew older, was seldom styled. On the other hand, Tesla, over six feet tall, cut an attractive, well-dressed, and even dashing, figure. When he was a young man, his dark hair and neatly trimmed mustache seemed to match his penetrating eyes and angular features. After he became a cult hero to many Americans in the 1930s, Tesla was caricatured in film in the Flash Gordon serials both as Dr. Zarkoff and Emperor Ming the Merciless of Mongo, who aimed his death ray at the Earth. Tesla-type mad scientists have also turned up as cartoon characters in comics and graphic novels. Today's stereotype of the mad scientist is still based on the popular conception of Nikola Tesla.

Edison and Tesla were also as different as different could be in terms of their temperament and method of approach to design and invention. Tesla was a visionary who, in an intuitive blink of insight, could solve problems almost instantly, while Edison depended on experiments that focused on trial and error. While Tesla spoke several languages, Edison was not as well-traveled as the brilliant Serbian. Tesla was impatient and, as some called him, "high strung." Edison, however, was the cleverer inventor and marketing expert, not only because he understood his market, but also because he had experience in practical inventions from the time he was a child.

AC VERSUS DC

When Tesla got his opportunity to meet Edison, he hoped to convince him that alternating current (AC) was superior to direct current (DC). Edison's direct current generation had become the standard for American electric power distribution by the late nineteenth century, essentially because it was the only type of electrical generation available. But Tesla, maybe naively, thought he could convince Edison to change all that. Tesla had been working on an AC motor that he promised would be better than Edison's DC version because it was more efficient and generated a more powerful electrical supply across a distance. It generated at a higher frequency than direct current and it had a greater potential, the result of moving the charge closer to a like charge and then away from that like charge. The result is that there was greater potential energy which, Tesla showed, meant that the current in a circuit could travel a longer distance without significantly degrading. That made it superior to direct-current generation, which required multiple sites of amplification and regeneration.

But this was a challenge that Edison was not prepared to accept, believing that his direct-current generators were not only more reliable, but were better able, through a system of substation relays, to deliver electricity at a lower, and safer, voltage. Besides, Edison had already invested heavily in direct current, had built an infrastructure, and was not about to throw out what he had been working on for over a decade. Edison was, by this point, the corporate industrialist protecting his technology.

Because Edison had no interest in Tesla's plan for alternating current, what began became known as the "War of the Currents," contested by Edison's General Electric Company and Tesla's George Westinghouse. Edison would stubbornly hold on to direct current, while Tesla did all but plead for alternating current. Because direct current travels in only one direction, it is limited to roughly a square mile from the power plant, after which it drops off considerably. Alternating current, on the other hand, constantly reverses its direction, thereby providing the electrical supplier a much greater range. Edison's earliest power-generating station was on Pearl Street in what today is lower Manhattan. But it only serviced a small area. Therefore, a city as large as New York, even in the 1890s, would have required a power-generating station supplying direct current in every neighborhood, a highly expensive proposition not only in terms of building costs, but in terms of the amount of real estate required. Worse, direct current also meant that supplying rural areas with electricity would be almost impossible. Those rich enough to afford it could license from Edison the right to construct their own local power plant, also a formidable obstacle to a nationwide power grid. Alternating current is not restricted by such boundaries and can be conducted affordably over larger areas because it has greater potential (power) than direct current.

Why did Edison obstinately refuse to consider alternating current? By embracing Tesla's AC, he could have revolutionized the business of municipal power supply decades before it was adopted.

But he either truly believed that direct current was better or he could not bring himself to admit that he was wrong. It was no secret that Edison was stubborn and found it difficult to change his mind once he became committed to something.

While Tesla was employed by Edison, at a relatively small salary, the Wizard of Menlo Park promised Tesla $50,000 if he could overhaul the jumbo generators at the Edison plant on Pearl Street in Manhattan. Tesla completed the work satisfactorily, but when he asked Edison for the money he was promised, he was turned down. Edison told Tesla he'd been joking when he made the offer, but Tesla did not see the humor. Furious, Tesla quit working for Edison's company.

Not long after, Tesla came to the attention of another entrepreneur and inventor, George Westinghouse, who had developed the railroad air brake. Westinghouse hoped to develop a competing power system, but soon realized Edison had the market cornered on direct current. Instead Westinghouse turned to AC, the system Tesla hoped to see put in practice. Edison called Westinghouse "crazy" and Tesla "impractical."

When his first electrical system was implemented in the town of Great Barrington, Massachusetts, Westinghouse employed a large and skilled sales force to encourage residents and businesses to replace gas with AC. Edison's direct current did not sell as well. Each company, Edison and Westinghouse, touted its service as the best, just as Edison had previously maintained that electricity in one's home, office, or factory was safer than using gas to light a room or building. In time, electricity all but replaced gaslight around the world. And because of Westinghouse's marketing skills and ability to demonstrate the financial benefits of AC versus DC, alternating current came to be seen as more efficient in the battle of the currents. But Edison was not a man to be trifled with and would not take impending defeat in a sportsmanlike manner.

The battle of the currents became so heated that Edison proposed using the Tesla/Westinghouse AC motors to create the electric chair. This, at a time when New York State was looking to replace hanging and find a more humane way to execute its condemned prisoners. This was not an idea wholly conceived by Edison, because news reports had carried stories of people who had become electrocuted by touching the hot AC wires bringing current to railroad trains. Thus, AC was known to be dangerous. But just how dangerous would be a demonstration that Edison believed would turn the world away from AC and towards his DC generators.

THE ELECTRIC CHAIR

Edison next hired Harold Pitney Brown to build a device that would prove that AC was more dangerous than DC. To test the device, Edison began experimenting by jolting animals with lethally high current. In fact, his Menlo Park invention laboratory soon became an execution chamber where rodents, shelter dogs destined to be euthanized, and even a circus elephant were electrocuted by alternating current. At this point, Edison knew it could be done. All that remained was to construct a chair, which was the job assigned to Harold Brown. Of course, Edison was not really looking for a more humane way to execute prisoners. He was really looking for a way to demonstrate that AC was much more dangerous than DC.

To do this, after Brown had constructed the chair, Edison bought and installed Westinghouse AC motors in New York's Auburn Prison where convicted killer William Kemmler was waiting on death row for his execution. George Westinghouse himself sought to intervene in the first execution, claiming that using electric current to kill a human being was not only painful but inhumane punishment and a violation of the Eighth Amendment. Edison prevailed despite Westinghouse's protestations. Then, in a scene that might have come right out of the motion picture *The Green Mile*, when the condemned William Kemmler was seated in the chair and the current turned on, he convulsed violently with tremors as the searing voltage passed through him, creating sparks and smoke. He did not die immediately, so the current had to be applied again, at which point Kemmler's heart stopped and he was declared dead. Edison had made his point. The execution was such a gruesome event that the newspapers reporting it said that Kemmler had been "Westinghoused." That was the length to which Edison went in order to win the battle of the currents, a battle he ultimately lost.

Edison was being disingenuous in his criticism of Tesla's alternating current. Not only did Tesla understand that the high frequency and potential of AC was deadly when carried by heavy supply lines, he also understood how to step down the voltage when it was transmitted to private homes and businesses. To do this, Tesla patented his transformer (US Patent Office number 593,138, dated November 2, 1897), a mechanism that reduced voltage going to a residence so that it would be safe in case someone came in direct contact with it. Tesla's transformer made AC current safe even though Edison claimed that it was lethally dangerous.

TESLA'S INVENTIONS FOR THE FUTURE

Tesla moved forward to invent new machines and instruments, focusing his attention on ways to eliminate the expensive and cumbersome cables and overhead wires. He invented the wireless transmission of electricity and a turbine-like device that generated very high frequency current. He invented what today is known as the

Tesla coil, a transformer circuit used to produce high-voltage, low-current, high-frequency AC electricity. It not only could produce great power, but had what Tesla claimed was "intelligence." Movie fans may remember Tesla coils harnessing and amplifying lightning to give life through electricity to Boris Karloff and Lon Chaney, Jr., actors who both portrayed the Frankenstein monster during the 1930s and '40s. In a May 1899 interview Tesla gave to *Pearson Magazine*, he tried to explain an invention that would harness the sun's power to drive a generator that would produce AC. As Tesla explained it:

> This is the experimental model of the apparatus with which I hope someday to so harness the rays of the sun that that heavenly body will operate every machine in our factories, propel every train and carriage in our streets, and do all the cooking in our homes, as well as furnish all the light that man may need by night as well as by day. It will, in short, replace all wood and coal as a producer of motive power and heat and electric lighting.[45]

This idea of converting sunlight into steam so as to drive generators was an amazing concept for its time. Even though it was, as Tesla's interviewer admitted, a bold idea, it was such a simple concept that even lay people, non-engineering types, and schoolchildren could understand it. It consists of concentrating and amplifying the heat of the sun on one spot (the glass cylinder) by the series of complicated mirrors and magnifying glasses until the resulting heat is most intense.

---

45 M'Govern, Chauncy Montgomery, "The New Wizard of the West," *Pearson Magazine*, May, 1899.

"This manufactured heat is directed upon the cylinder filled with water chemically prepared so that in a short time the water has evaporated into steam and has passed from the cylinder through a pipe and into another chamber. In the latter place, this sun-made steam piped into a regular steam engine of ordinary construction, the horse-power of which will be determined by the size of the apparatus by which the sun generates steam in that spot. This steam engine is used to generate electricity. And this electricity can be either used at once or else stored up in storage batteries to be used on days when there is no sunlight."[46]

You can imagine what the impact of Tesla's sunlight conversion idea would have been for the twentieth century. By 1900, coal was the principal fuel for the steam engines that drove electric turbines. By the 1950s, oil and then natural gas became the principal fuels, and then, for a time, nuclear power. Today, our coal reserves are running out and hydraulic fracturing, or "fracking," for natural gas provides a much cheaper and efficient way to generate fuel, albeit not without its own carbon overhead. With the exception of nuclear power, which, of course, presents its own liabilities, coal, oil, and gas are all pollutants. But converting the sun's rays, via a system of mirrors, into a heat source to heat water to steam to drive turbines would have transformed the twentieth century from generating greenhouse gases to having almost pollution-free electric generators. In addition, since a solar-generating plant does not rely on the purchase of fuel, the cost of electricity would have been remarkably low. It conceivably could have transformed society and brought power to Third World countries in such a way that the politics of oil might have never destabilized much of the world.

---

46 Ibid.

As Tesla prophesied, "In this way electricity will be so cheapened, that it will be possible for the poorest factory-owner to use it as a power at a smaller cost than steam. Electricity will in this way supplant steam as a motive power on all railways and—in the shape of storage batteries—on all water vessels. And the humblest citizen will profit by the new system of producing electricity; for he can have it in his home to do all his cooking and lighting and heating. And it will be even cheaper for him than coal, wood, or petroleum."[47]

When one looks at Tesla's prediction in the face of what was driving Edison's businesses, it's clear to see how that challenged the very industrial model Edison had spent years trying to build. Not only would this have been a terrible blow to Edison's ego, that a foreign-born eccentric engineer could supplant him, but given the next-to-nothing cost of generating solar power, it would have driven his power-generating companies out of business. Thus, by the beginning of the twentieth century, Edison and Tesla were rivals for the heart and soul of America's power and fuel industries. It is no stretch of the imagination to see why Edison would have gone to such great lengths to debunk not only the theory of alternating current, but to debunk Tesla as a man of science.

HYDROELECTRIC POWER AND WIRELESS ELECTRICITY

Exploiting solar power to convert amplified heat into electricity would not be the last topic over which Tesla and Edison would clash regarding free versus metered energy. The next conflict would come to a head when Tesla sought to beam energy wirelessly to

---

47 Ibid.

power cities, thus eliminating the need for a hard-wired power grid. In the immediate aftermath of his experiments with solar power, however, Tesla turned to water-driven turbines to generate electricity. And he used the power of Niagara Falls.

One of Tesla's most amazing triumphs was to harness the tremendous force of Niagara Falls, whose water power he used to drive a hydroelectric generator. He was so successful in this that in 1895 Westinghouse built a gigantic hydroelectric plant, employing Tesla's alternating current system. This was also a blow to Edison's investment in delivering direct current to homes and businesses. The irony of the Tesla and Edison feud is that it spread to the two companies, Westinghouse and General Electric. More ironic is that Westinghouse became part of the Viacom/Paramount organization, a CBS corporate conglomerate, while GE wound up being a part of the Universal/NBC conglomerate. In a sense, the Tesla and Edison rivalry stretched across the centuries, as the television networks battled over ratings and the studios vied for theater goers.

THE CHICAGO WORLD'S FAIR

Perhaps the most serious dispute in the war of the currents between Edison and Tesla came to a head at the 1893 Chicago World's Fair. The exposition, which commemorated the 400th anniversary of Columbus's discovery of the New World, had let out bids for building the lighting and power system. It was not only a potentially lucrative contract; the importance of the exposition was such that whoever won the contract would have had a proof of concept design and implementation to show off to the world. How could other cities, industrial campuses, and local municipalities looking to supply power and light to their residents fail to be impressed by any display at the 1893 Chicago World's Fair?

Both General Electric and Westinghouse sought to win the contract to light the fair and put in competing bids although Edison, with J.P. Morgan investing and General Electric contracting, did not know there were any competitors. Perhaps thinking that they were all alone in the bidding and maybe even thinking that only they had the technology to install the power supply, General Electric put in a bid for $1.8 million. Maybe they were confident that even if the bid were considered slightly too high, they would have room to negotiate down, given that they probably overestimated their own costs. The bid was rejected by the contracting agency and GE was forced back to the accounting sheets. They soon came back with a counteroffer, a bid of approximately $550,000. Now they were certain of winning the contract. Who could turn that down?

It might have been a good bid for General Electric had they been the sole bidder. But they did not know that Westinghouse had also assembled a proposal to light the fair at a price even lower than GE's, at just under $400,000, Westinghouse said they could light the entire fair, beating General Electric by $150,000. Using Tesla's AC technology, Westinghouse proceeded to successfully light the fair and, in so doing, demonstrated that alternating current could be used in municipalities across the nation. Alternating current eventually became the power grid for America.

TESLA, THE COURT-DESIGNATED INVENTOR OF RADIO TRANSMISSION?

Tesla also experimented with "electromagnetic waves" that were the foundation of radio. In fact, although Marconi is reputed to have generated the first radio signals across the Atlantic, Tesla claimed that he was the first to invent the wireless transmission of radio waves. That claim eventually went to the United

States Patent Office and ultimately to the United States Supreme Court. The battle between Tesla and Marconi seemed to have been resolved by the patent office, which awarded the patent to Marconi, but Tesla maintained his challenge in the courts. Then, at the outset of World War I, Marconi sued the United States government for a patent infringement because the US was using radio transmissions for military communications. This did not sit well with the military and they supported Tesla's claim in court for the patent for radio transmission. In 1943 the Supreme Court extinguished Marconi's claim for patent infringement and his claims for royalties. In *Marconi Wireless Tel. Co. v. United States*, 320 U.S. 1 (1943), the court held, "The broad claims of the Marconi Patent No. 763,772, for improvements in apparatus for wireless telegraphy—briefly, for a structure and arrangement of four high-frequency circuits with means of independently adjusting each so that all four may be brought into electrical resonance with one another—held invalid because anticipated." Tesla never had a chance to celebrate his victory, having passed away six months earlier.

Ironically, it was Edison's company, General Electric, which had acquired the rights to Marconi's radio patent. They in turn had exploited it in the creation of their motion picture company, Radio Corporation of America, or RKO, owned in part by Joseph P. Kennedy, JFK's father, who was instrumental in using Edison's and Marconi's technology to develop one of the first studios to make talking pictures. Marconi's employee, David Sarnoff, worked on wiring the first RKO studios with sound before he founded the National Broadcasting Company, now a part of Universal, General Electric, and Comcast. When the Court ruled against Marconi, it was as if Tesla had defeated Edison again. Thus, in popular lore today, it is Tesla who prevailed as a matter of law even if not of science, defeating not only Marconi, but Edison and Joe Kennedy.

TESLA AND REMOTE RADIO CONTROL

According to author Margaret Cheney, Tesla also had a grasp of the underlying theories of generating X-rays, of the electron microscope, and of using radio-controlled devices with torpedoes and even missiles.[48] It may seem strange, when considering the nature of Tesla's far-ranging inventions, that although he did not seem to understand the science of his visionary experiences, there was no question that he had an abundance of accurate paranormal visions.

Tesla wrote in his diary, "During my boyhood I had suffered from a peculiar affliction due to the appearance of images, which were often accompanied by strong flashes of light. . . . Then I began to take mental excursions. . . .

"This I did constantly until I was seventeen, when my thoughts turned seriously to invention. Then I found I could visualize with the greatest facility." Tesla went on to explain how he did not need to concentrate, he could simply visualize a future invention. "The inventions I have conceived in this way have always worked. In thirty years there has not been a single exception," he wrote. "Before I put a sketch on paper, the whole is worked out mentally."[49]

Tesla also claimed that he had discovered "an inexhaustible source of energy that could be transmitted anywhere in the world without wires or loss of power." He accurately predicted the future, stating that someday "it will possible for nations to fight without

---

48 Cheney, Margaret, *Tesla, A Man Out of Time*, (New York: Dorset Press, 1989).

49 All quotes from Tesla's journals are from Birnes William J. ed., *Tesla Illuminated*, (Los Angeles: Filament Books, 2007).

armies or guns . . . with weapons far more terrible." Tesla discovered a "protective radiation principle," popularly called a "death ray."[50]

TESLA AND EXTRATERRESTRIAL LIFE

Arguably, Tesla's most controversial area of interest was in extraterrestrial life, an area where he also competed with Edison. During 1895 and 1896, he considered that there might be life on other planets. He wrote, "If there are intelligent inhabitants of Mars or any other planet, it seems to me that we can do something to attract their attention. . . ."[51]

He also delved into the language of Theosophy or Hindu metaphysics. Tesla's states of higher consciousness achieved by "intense concentration and a celibate life" seem like the Hindu ideas of cosmic energy in the universe. It also suggests that Tesla read or studied the writings of Madame Helena Petrovna Blavatsky, the founder of Theosophy. He also had a rudimentary knowledge of the Vedic texts, which describe the Hindu deities battling each other from flying saucer-type devices and firing death rays at each other.

Tesla also theorized the principles of what he called "wireless telegraphy," essentially radio, and "visual wireless telegraphy," television. The way he described it was through the analogy of a balloon filled with water with piston-like devices at both ends. Suppose, Tesla said, that at one end of the water-filled balloon, you pressed a piston into the liquid so as to expand the bag, while at the other end a piston withdrew the expanded liquid and transmitted it. Now imagine that the Earth's atmosphere is the balloon and a transmitter,

---

50 Ibid.

51 Swartz, Tim, ed. *The Lost Journals of Nikola Tesla*, (New York: Inner Light-Global Communications, 2000).

say a telephone device, is at one end and a movie-like screen is at the other end. What you transmit, visual and audio signals through the telephone, now come out the other end. In essence, not only was Tesla describing what would become television, he was also describing what would become today FaceTime and Skype.

THE TESLA WIRELESS GRID

In the Sunday edition of the *New York Journal* on August 8, 1897, reporter Julius Chambers described an interview with Tesla in which the inventor explained how he would be able to draw current from existing hydroelectric generators, amplify that current to higher frequencies and potentials, and then broadcast that current wirelessly to receiving stations around the world for sub distribution. This was Tesla's vision of free electrical power, something that would have directly challenged the Edison General Electric model of metered electrical supply for which customers would pay a monthly fee. This is the model in use today for municipal power suppliers. Ironically, Tesla's idea of free energy distribution also posed a direct challenge to his former investor, George Westinghouse, who financed Tesla's alternating current distribution system.

As early as 1890, even before he constructed his Wardenclyffe tower in Shoreham, New York, Tesla explained his invention of wireless transmission to his interviewer at *Pearson* magazine. He said that just as easy as it was to transmit electrical power via cables strung about on poles, imagine if the Earth itself were the cable and that all the construction expense of building the grid were eliminated. Imagine if power generating and transmitting stations could be constructed near every hydroelectric facility. Think of the cost savings, think of simply using the Earth's own natural resources instead of manufactured lines and relays. Simply speaking, the atmosphere takes the place of the cables, with the electrical signal, the flow of electrons, traveling freely through the air. Tesla conceived an array of towers around the country from which he would suspend balloons high in the atmosphere. The balloons would act as transmitters in the rarified atmosphere, where signals would be transmitted with much less resistance. As for receiving stations, Tesla proposed to erect the same structures with the balloons equipped with devices to absorb the free-floating electrical signals, then transmitting them down though the towers via cables. A system of wires and cables, Tesla told his interviewer, would convey the power supply locally. Thus, imagine a grid much like Edison's with the exception that the atmosphere would replace the long cable supply lines, with local stations receiving antennae.

Tesla was passionate about his discoveries into the realm of wireless transmission, claiming that he had harnessed the power of the Earth's magnetosphere as the distribution mechanism. He claimed that by projecting a wireless beam of electrons, literally a particle beam weapon well ahead of its time, at any form of receiving station, that beam would power that station. Hence, he could even light ships at sea, far from land, simply by directing his beam in that direction. In fact, Tesla said, his particle-beam distribution system was so powerful that he could project an electrical disturbance toward Mars and, assuming there were life forms on distant planets capable of receiving his particle beams, he could communicate with them. Thus, by the beginning of the twentieth century, Tesla claimed that he had invented a form of wireless distribution that was, we know now, one hundred years ahead of its time.

Tesla refined his idea by 1911, telling the *New York Herald* that his plan was nothing less than turning the entire planet into a giant worldwide dynamo of electrical generating power.[52] Of course, his Tesla towers sparked and crackled, he said, transmitting a form of lightning through the air. And, yes, this was very high current, higher than the high-volume main electrical supply trunk like those we see today hanging from power poles across rural America. It was dangerous, and in some areas of government there were complaints that a misdirected wireless strike might have the same effect as the electric chair.

The main transmission tower was erected in 1897 in Shoreham, Long Island, in New York State, with the goal of transmitting power across the Atlantic. It had been financed, ironically, by J. P. Morgan, who believed he could monetize wireless transmission. But the tower also served as the proof of concept that signals could be transmitted without wires. While the dispute over Tesla's and Marconi's rights to patents would surface in court prior to the US entry into World War I and again in World War II, the United States military was also wary about the tower. The military believed that German agents might try to use the tower to transmit information about the American military presence in the Atlantic. In fact, the military informed the press that they had seen "suspicious" individuals surveying the tower property, and believed them to be German spies. Because of these concerns, the military blew up the base of the tower, which then collapsed. During World War I, German naval vessels had deposited spies on the eastern edge of Long Island, near Montauk, and the spies had made their way west toward New York City. They were eventually captured and held by federal agents and the military.

Newspapers at the time reported that Tesla was regarded as a genius, albeit an eccentric scientist, by the American public. The

---

52 "Tesla's New Monarch of Machines" *New York Herald*, (October 15, 1911).

public, while wary of lightning strikes on power-receiving stations, was also excited by Tesla's promise of unlimited power transmitted through the air. By the outset of World War I, Tesla had become a symbol of American inventiveness, even more so than Edison, the practical man of science. While Edison was making motion pictures for entertainment, improving the telephone for mass communication, and working in the financial services and stock trading industries, Tesla was building death rays and particle-beam transmitters that, he said, would enable people on earth to talk to life forms on other planets. This was not just a contrast of ideas, it was a contrast of cultures and belief systems. But, we argue, there is no question that the Edison–Tesla rivalry must have prompted Edison to explore new ventures at the envelope of the consumer market.

Tesla's demonstration of the efficacy of wireless transmission, not just of radio waves, but of electrical power itself, meant he had managed to control particles—electrons—which, because of their collective charge, seemed to have some cohesiveness. Just three years after the destruction of Tesla's Long Island tower, Edison ventured into the same area of receiving wireless transmission, only this time not from radio or power transmitters, but from free-floating bundles of what he called "life units." These were electrically charged particles that he believed might carry the sentience that was embodied in their human form. This was the underlying premise of his spirit phone, that it could detect electrically charged life units. And there was no question that his rivalry with Tesla is what inspired Edison to make one last invention, this time into the world of the spirits and the departed, a place that Tesla had already visited—if not in person, then in books.

TESLA'S LATER YEARS

Nikola Tesla was eccentric as well as remarkably psychic and highly intelligent—so intelligent, in fact that he was awarded the Nobel Prize in physics in 1912. He turned it down because it would have to be shared with his bitter rival Thomas Edison. The award went instead to Swedish scientist Nils Gustaf Dalén.

Tesla fell in love only once in his life, and that was when he was a young man. When the relationship did not work out, he never again became seriously or romantically involved, nor did he ever marry. He lived in solitude with his own thoughts and his own inventions and, some said, devolved into a form of solipsistic dementia.

In his last years, Tesla lived alone in the New Yorker hotel in a single room in virtual poverty, an ironic twist of fate for a man in whose work financiers like Westinghouse and J. P. Morgan had invested millions. It was a sad end to a brilliant man's life and career. Shortly after his death in January 1943, FBI agents opened a safe in Tesla's room and carted away its contents. One theory is that government operatives took papers that some think held details of a "secret invention" that may have been useful in

STORY OF TESLA'S ANTIGRAVITY NOTES

- Receives grant for antigravity from Soviet Union at outbreak of World War II
- 1943: Hotel room raided after his death by FBI
- 1944: All notes seized by Office of Alien Properties
- 1946–1950: Tesla's notes returned to Yugoslavia after World War II for a Tesla memorial museum
- 1947–1948: Antigravity notes not returned but sent to General Nathan Twining at Wright Field, where Roswell crash debris was sent
- 1950s: Theory of Tesla's antigravity device: ultra-sound waves levitate an object

warfare. They took the notes of Tesla's secret invention, plans and schematics for which the Soviet Union had awarded him a development grant, and turned them over to General Nathan Twining at Wright Field, Ohio. There, just four years after Tesla's death, the debris was taken from a crash near Roswell, New Mexico. Visitors to the scene, including President Harry Truman, described the craft as extraterrestrial, and stated that the alien life forms that navigated it were also taken to Twining's Air Materiél command for analysis. What was Tesla's collection of notes on his plans for a secret weapon? It was antigravity, the very mechanism that seems to have allowed the Roswell flying saucer to skip through the atmosphere with no apparent traditional form of propulsion.

In recent years, there has been a renewed interest in Tesla with all his complexities and peculiarities. Perhaps he will finally get the recognition he has largely been denied.

CHAPTER 12

Edison, Tesla, Robots, and Artificial Intelligence

Both Edison and Tesla had articulated exotic and unconventional theories about artificial intelligence. Edison's belief (shared with Tesla) that everything was alive helped form the basis of his theories for the development of the spirit phone. Edison wrote extensively about

TIMELINE

January 25, 1921: Premiere of Karel Čapek's Czech language play *RUR* (*Rossumovi Univerzální Roboti* or *Rossum's Universal Robots*)

what he termed "life units," primary elements of life, particles of electrical energy, which were the basis of all existence. Even things that seemed inorganic, such as rocks, were infused with life units that gave them substance. In this regard, Edison quite presciently foresaw the modern theories of Gaia, the living Earth. Edison was giving a technological explanation to what primitive people said was living nature, a world motivated by deities that were responsible for how things came into being. This may seem strange for a man who was

151

imbued with devout Methodist beliefs, but Edison was remarkable in that he had the ability to assimilate different strands of thought and belief, faith and science, into an aggregate vision of working reality. This was how he managed to combine scientific theory and spiritualism into a design and process for his spirit phone.

Edison's theories of intelligence abstracted from human form diverged from traditional theories of existence in the way life units organized themselves. Edison believed that if these electrically charged units existed after the death of the body, they would be attracted to each other—a precursor to Einstein's theories of quantum entanglement and spooky action at a distance—in such a way as to operate with a form of abstract intelligence. This intelligence, Edison believed, could be identified and pinged by a device that registered these floating but cohesive life units.

MODERN QUANTUM THEORY AND EDISON'S THEORIES OF CONSCIOUSNESS

It has taken almost one hundred years, but theoretical physicist Sir Roger Penrose, basing his argument in part on near-death experiences, has argued that the microtubules in human cells can exist after the death of the body as quantum information. They exist as separate, but cohesive, units apart from the body, entangled with each other according to Einstein's theory of spooky attraction at a distance. Just as Edison predicted, Penrose argues that the cohesive quantum units remain entangled, providing a form of consciousness.[53] They can even return to the body cells if the body itself comes back to life. This, he argues, is his proof of existence at a quantum conscious

---

53 Harrison, George, "Soul Searching, Researchers Claim that Humans have Souls which can Live on After Death," *The Sun*, (November 5, 2016).

level after the body's death. In this theory, Penrose is supported by physicists at the Max Planck Institute for Physics in Munich.

TESLA'S THEORIES

Tesla went a step further than Edison. Like Edison, he believed that all existence, organic and inorganic, is imbued with life. He based his theory on what he stated as a fact that everything we see around us is capable of responding to irritant stimuli. According to Tesla, as quoted in the *New York American* on February 7, 1915, "Even matter called inorganic, believed to be dead, responds to irritants and gives unmistakable evidence of a living principle within. Everything that exists, organic or inorganic, animated or inert, is susceptible to stimulus from the outside." Moreover, possibly because he believed that even the inanimate was alive, he believed that he could provide objects that were inanimate with a kind of self-directed consciousness. In other words, he could devise an automaton that would act of its own volition to follow instructions given to it by human beings.

Tesla's theories were truly advanced. In an imaginary interview written by Marc J. Seifer and Michael Behar in the October 1998, issue of *Wired* Magazine, based on Tesla's own words, Tesla says, "My plan was to construct an automaton which would have its 'own mind,' and by this I mean it would be able, independent of any operator, in response to external influences affecting its sensitive organs, to perform a great variety of acts and operations as if it had intelligence. It will be able to obey orders given far in advance, it will be capable of distinguishing between what it ought and ought not to do and of recording impressions which will definitely affect its subsequent actions. Further I do not believe that intelligence is artificial, but rather a property of matter."

Even before the 1921 performance of Karel Čapek's *RUR*, Tesla had conceived of devices that could be instructed to perform certain tasks and use internal logic to make decisions based on programmed instructions. He designed weapons such as robotic ships and torpedoes that could distinguish between friend and foe based upon precisely formulated algorithms and then make the appropriate tactical moves. He said, "I conceived the idea of constructing such a machine, which would mechanically represent me and which would respond as I do myself, but of course in a much more primitive manner, to external influences. Whether the automaton be of flesh and bone, or wood and steel, it mattered little, provided it could undertake all the duties required of it like an intelligent being. With machines to do the work, man will be that much more free to increase his knowledge and productivity and thereby advance the planet."[54]

Even as early as the turn of the century, Tesla's theories of artificial intelligence were challenging both the religious and cultural trends of the time. Like most of Tesla's theories, including wireless transmission and using solar power as a fuel to generate electricity, his theories of artificial intelligence or AI have not only come to be, but are engendering scrutiny from the people at the heart of its development. Scientists and product developers such as physicist Stephen Hawking, Apple co-founder Steve Wozniak, Tesla Motors project leader Elon Musk, and computer scientist Ray Kurzweil, all of whom have predicted the ultimate creation of an artificially intelligent machine, have given dire warnings about the dark side of AI, in particular how an AI machine will deal with humanity. For example, according to Christopher Mims in the Wall Street Journal (June 25, 2017), within ten years, an AI incarnation

---

54 Seifer, Marc J. and Michael Behar, *Wired* (October, 1998).

of Apple's Siri will not only will replace the smartphone as we know it, but will be integrated into the glasses and clothing we wear and devices in our ears and will adapt itself to our own personalities and proclivities on a minute-by-minute basis.

The issue of the relationship between an artificially intelligent robot and human beings was raised by Isaac Asimov in the science fiction universe he created in his series *I, Robot*, in which he stated the fundamental laws of robotics (a word he created), which, verbatim, were: "(1) A robot may not injure a human being or, (2) through inaction, allow a human being to come to harm. And (3) a robot must obey orders given it by human beings except where such orders would conflict with the First Law, and a robot must protect its own existence as long as such protection does not conflict with the First or Second Law."[55] Actually these three laws were first articulated in the Asimov 1942 short story *Runaround*, published in the magazine *Astounding Science Fiction* and republished eight years later in *I, Robot*. In its modern incarnations, these laws of robotics were again promulgated in the science fiction movie *Forbidden Planet* (1956) and in the personification of the *Star Trek: The Next Generation* android "Data."

WESTWORLD AND THE ORIGIN OF ARTIFICIAL CONSCIOUSNESS IN THE ROBOTIC BRAIN

Author Michael Crichton's *Westworld* took the distinction between artificially intelligent robots and humans to a new level when the robots become resistant to their programming.[56] After the *Westworld* feature film, which Michael Crichton directed, came the HBO television series *Westworld* by TV producer J. J. Abrams.

---

55 Asimov, Isaac, *I, Robot*, (New York: Bantam, 1950).

56 Crichton, Michael, *Westworld*, (New York: Bantam, 1974).

These storylines revolved around the birth of consciousness in androids after a rogue designer inserted the propensity for memories to feed into their reactions to their daily experiences and carry over from one programmed story line to the next. As a result, the bleed from memories of prior relationships, of suffering at the hands of others, or of death, albeit artificial, informs their processing of sensory inputs, thus rendering them susceptible to human reactions and not just pre-programmed responses.

The philosophy behind Abrams's incarnation of Michael Crichton's *Westworld* is based in part upon the important work of the late Princeton psychologist Professor Julian Jaynes, whose theses are embodied in his seminal work *The Origin of Consciousness in the Breakdown of the Bicameral Mind.*[57] In this, he argued that what we call self-awareness was the result of messaging from the subordinate hemisphere of the brain to the dominant hemisphere, albeit that the messaging itself was beneath the level of conscious awareness. It wasn't until the actual formation of the hardwired nature of language, communicated between the hemispheres through the bundle of neurons known as the corpus callosum, that the brain became asymmetrical and the formation of logic and language began.

It was this force-feeding of the conscious by the subordinate conscious that inspired Freud's construct of the way memories and emotions operate on the conscious mind, and what the CIA experimented with in Canada in the 1950s under its MK-ULTRA program. As related in Chapter 7, they were doing it to uncover

---

57 Jaynes, Julian, *The Origin of Consciousness in the Breakdown of the Bicameral Mind,* (Boston: Houghton Mifflin, 1976).

the memories of deep cover programmed Soviet spies sent into the US with false American identities after World War II and the Korean War. Modern popular psychology titles from the end of the last century, such as Eric Berne's *The Games People Play* and Dr. Thomas A. Harris's *I'm OK, You're OK*, came out of this program from the US government's sponsored research of physicians Wilder Penfield and Ewan Cameron, both CIA contractors in Canada.

Their work, although originally intended to uncover the existence of Communist spies and saboteurs during the Cold War, gave psychologists new insights into human motivations and the nature of sentience and consciousness. This was at the very dawn of the age of computers, whose programmers sought to mimic the human condition through artificial intelligence and algorithms that can set in motion a form of self-determination. This was one of the themes of *Westworld*.

THE PROMISE AND THE THREAT OF AI

The rules of robotics notwithstanding, the reality is much starker when we realize that not only are military scientists and weapons developers finally exploring Tesla's theories of robotic soldiers, even if these soldiers exist only in computers as aggregations of electrons absent physical bodies. For example, in Joel Achenbach's December 27, 2015, article in the *Washington Post*, he poses a question that perplexes futurists dealing with advanced computer intelligence technologies: "What would happen to humanity if AI machines are allowed to self-develop?" Futurist Charles Ostman, who has appeared on late-night radio talk shows such as *Coast to Coast AM* and *Future Theater*, has also argued that AI developers are confronted with the issue of programming AI devices in an entirely new way. For example, in the television science fiction series *Extant*, a supercomputer whose primary mission is to defend the planet assumes a consciousness all its own and determines that the real threats to Planet Earth are the human beings who inhabit it, pollute it, desecrate it, and kill off other species. In response, the computer's algorithm sets about to solve the problem of making human beings less of a malignant species. Its solution: eliminate humans altogether, replacing them with androids who take their instructions directly from the supercomputer. This is the nightmare scenario Stephen Hawking, Steve Wozniak, and Ray Kurzweil warn us about, a struggle for survival between the Carbons (us) and the Silicons (artificial life forms). And so did Michael Crichton.

At the turn of the twentieth century, Tesla argued that a pre-programmed robotic soldier, or a torpedo, or a surface vessel operating under a set of preprogrammed instructions, could make its own decisions regarding where and when to attack. The United States Navy, and, we presume, the Russians, are already testing drone minisubmarines preprogrammed to carry out specific missions. Tesla even demonstrated his radio-controlled boat in New York City, but when he tried to sell his ideas about advanced automated weaponry to the Navy at the outset of World War I, it was Thomas Edison, then sitting on the Naval Consulting Board, who advised the Navy to reject Tesla's plans. The concept of a robot-guided torpedo did not fade away with Edison's turning Tesla down, however. It resurfaced over thirty years later when movie star Hedy Lamarr applied for a patent to make Tesla's idea into a workable weapon by developing with composer George Antheil what today is referred to as "spread-spectrum technology."

HEDY LAMARR AND THE SPREAD-SPECTRUM TECHNOLOGY

Tesla's ideas for radio-controlled weapons, although far ahead of their time, were also subject to a singular flaw in their inception: jamming. Think of someone keying a microphone on a CB radio where the static-heavy carrier signal will block any other transmissions on that frequency. We've already seen examples of signal jamming and, worse, signal spoofing, when the Iranians on December 4, 2011, commandeered the electronic signals guiding a United States military RQ-170 Sentinel unmanned aerial vehicle and ordered the device to land at an Iranian air base. They touted it as a major coup de guerre, demonstrating the vulnerability of American automated weapons systems. Edison had imagined this possibility when looking at Tesla's ideas. After all, if both sides were equipped with radio-directional antennas, the enemy might be able to home in on the radio frequency controlling the guided weapon, jam that frequency, and render it inoperable. How to prevent that?

The answer was provided by motion picture actress Hedy Lamarr and her partner, musical composer and co-inventor George Anthiel.[58] Together, they came up with the idea of spread spectrum frequency technology. Here's how it worked. Because all combatants were using radios at the outset of World War II, they understood that radio signals could be easily jammed, thereby preventing communication. For radio-guided weapons, the jamming potential posed a particularly difficult problem because the weapons could be rendered useless. This problem was potentially magnified because German U-boats stalking the North Atlantic convoys would have been able to jam incoming radio-guided allied torpedoes, making them miss their marks.

To solve this problem, Lamarr, a movie star by day and a genius inventor by night, and Anthiel, who, as a composer, certainly understood the nature of sound frequency, sought to camouflage the broadcast frequencies of radio-guided torpedoes by creating a spectrum of frequencies that the radio transmitter and receiver switched to automatically.[59] Only the transmitter and the receiver aboard the torpedo's guidance mechanism would be able to decrypt the changing frequencies along a closed circuit, which in

---

58 Delta, Yohana, "How Inventive 'Genius' Hedy Lamarr Became a Hollywood Tragedy," *Vanity Fair*, April 20, 2017, and Rhodes, Richard, *Hedy's Folly, The Life and Breakthrough Inventions of Hedy Lamarr, The Most Beautiful Woman in the World*, (New York, Vintage, 2011).

59 Delta, Yohana, Ibid.

turn would prevent any enemy jamming from taking place. In essence, the
enemy would not be able to determine the radio frequency to jam.

Lamarr and Anthiel filed a United States patent (#US2292387 A) for their
spread spectrum technology in June 1941. Their patent was published
in August 1942.[60] But, as Lamarr would complain late in her life, the
Navy never actually availed itself of this technology in World War II.
After fifteen years, the Sylvania Corporation registered its own patent for
spread spectrum technology, a technology that was implemented during
the Cuban Missile Crisis in October 1962, when Soviet submarines
approached the US naval blockade of Cuba.

Many people don't realize how close to a nuclear weapons exchange the
US and the USSR came. In fact, one Soviet submarine was ordered to
prepare its missiles for launch against the United States while an American
submarine shadowing the Soviet sub, silent, deep, lurking, had its torpedo
doors open. But the Russian weapons officers refused to comply with the
orders and the missiles were never launched. During the same few days,
US Air Force interceptors equipped with nuclear tipped air-to-air missiles
were dispatched to their failsafe points over the Bering Straits to await the
overwhelming numbers of Soviet bombers they expected to encounter.
It was an air battle that never took place because Soviet Premier Nikita
Khrushchev, according to former President Bill Clinton advisor Dick Mor-
ris, backed down from a confrontation after overplaying his guided missile
hand. In October 1962, despite Khrushchev's having told his Supreme
Soviet that the USSR possessed intercontinental ballistic missiles that could
reach the United States, the USSR did not actually have such weapons.
His backing down from a confrontation with President John F. Kennedy,
because he had overestimated his ability to dominate a heavily drugged
JFK at the Vienna Summit the year before, was his own death knell.[61]
After the Cuban missile crisis ended with JFK's removal of US intermedi-
ate-range ballistic missiles from Turkey and Russia's pulling its missiles out
of Cuba, Khrushchev lost his job as Soviet premier and party chairman.

The final footnote concerning the weapons deployed during that crisis is that
it was Tesla's concept of a radio-controlled torpedo and Lamarr's invention
of spread spectrum frequency technology that finally made their appearance
during the tense moments when both superpowers were girding for war.

---

60 Rhodes, Richard, Ibid.

61 Lertzman, Richard A. and William J. Birnes, *Dr. Feelgood*, (New York: Skyhorse, 2014).

As exotic as artificially intelligent computers and the weapons they control might be, what if, Futurist Ostman argued on the radio show *Future Theater*, an AI device were programmed to find a way to rid Earth of pollution, to bring the planet back to where it could sustain life without the extinction of species? In that instance, looking for pollutants, what if that algorithm focuses on the pollution caused by human beings and then decides on its own that humanity is the virus that's infected the planet? And what if it takes steps to eradicate that virus? After all, our planet had gone through climatological cycles for hundreds of millions of years and managed to keep on breeding life, therefore, why would humanity be the pollutant? Ostman argued that the overhead of human civilization, the deforestation of the planet, the growth of large urban centers, and the waste gasses spewed out by industry all contributed to making the climate extremes worse than if there were no human civilization. Accordingly, absent a "morality" algorithm, a set of limiters preventing the computer from harming its creators would turn AI computers into planet cleansers.

Computer scientist Ray Kurzweil has speculated about the moment one artificially intelligent system recognizes another, evolves into a self-replicating consciousness, then generates the code necessary to expand its awareness until it becomes conscious of itself as an existing entity. In that moment, which he calls a "singularity," how would that artificially intelligent device manifest itself? Would it announce its existence over a trillion telephones in a "Lawnmower Man" moment, declare its presence, start issuing orders to human beings, reach out to other digitally controlled devices such as power grids and dams, and ultimately follow Asimov's third law to protect itself?

Even more menacing is the fact that the military has taken advanced steps in creating AI-based weaponry. For example, in the computer guidance, tracking, and weapons systems of the F-35 joint services fighter/interceptor, a computer at the very edge of artificial intelligence can track multiple targets simultaneously, relay that targeting information to its pilot's helmet as well as to other pilots in the combat squadron, then issue firing instructions after identifying and locking on to the targets. All of this is accomplished at the speed of a computer, surpassing human reaction time. If this is the future, the robotic future is happening now as a weapon absent the prohibitions laid down by Asimov.

The technology supporting the F-22 fighter also borders on the edge of artificial intelligence. Onboard radar sensors are capable of gathering information about a target, making determinations about the target, and relaying that information to other pilots in the battle group or to flight controllers and targeting information processors on the ground. In mock air battles with the now obsolescent F-15, the F-22 scored a kill ratio of 180 kills to no losses.

ARTIFICIAL INTELLIGENCE SYSTEMS AND THE DOWNING OF TWA FLIGHT 800

Investigative journalist James Sanders[62] reported that on July 16, 1996, over the south shore of Long Island, New York, the United States Navy shot down a commercial Boeing 747 jumbo jet on its way to Paris from JFK International with 230 passengers and crew aboard. All perished when the plane exploded in midair and crashed into the ocean. After almost twenty years of investigation, Sanders

---

62 Sanders, James, *The Downing of TWA Flight 800*, (New York: Kensington, 1997; 2nd ed. 2013).

was able to determine what happened on that night. He wrote, and this was confirmed by official Navy documents, that there was a military exercise taking place that night in which a drone, probably simulating a cruise missile or anti-ship missile, was fired from the land over the ocean toward the south shore of Long Island. The mission was to shoot down that missile from a submerged Navy submarine inshore. This shallow-water exercise was designed to test the feasibility of a submarine defending itself during brown water operations. But the exercise turned out much worse than anyone could have imagined.

The exercise was a test of a new weapons system designed to take the fire control decisions out of the hands of human beings and place them in the hand of an AI-driven computer system, essentially a digital robot. It was a reaction to the July 3, 1988, shooting down of Iran Air Flight 655 by the Aegis missile destroyer USS *Vincennes* over the Straits of Hormuz. It occurred during an operation in which American naval vessels were escorting oil tankers through the straits in the face of Iran's threat to block the international seaway and prevent tankers from sailing toward their destinations.

The commanding officer of the *Vincennes*, Captain Will Rogers III, said he believed that Flight 655 was not a commercial airliner—it actually was an Airbus A300—but an F-14 Tomcat warplane on a mission against his warship. He said that he radioed the aircraft on a standard military hailing frequency to request identification. If Rogers had broadcast on a civilian frequency, he would have identified the aircraft as a commercial airliner. But he only tried to hail it on a military frequency and, getting no response, he misidentified it as a fighter bomber on an attack mission.

The vectors of the radar signatures were likely also confusing. Captain Rogers and his weapons officers believed that what they thought to be an F-14 was diving upon them for a bombing run. Instead, the plane was actually climbing out of a commercial airport using the proper flight lane in its ascent. The *Vincennes*'s radio warnings were also broadcast on military frequency, as opposed to commercial aviation, and thus the pilot did not respond. Consequently, even though other warships in the region believed Flight 655 to be no threat and did not turn their weapons on it, the *Vincennes* did and fired two anti-aircraft missiles, bringing the plane down and killing all 290 passengers aboard.

This was an event so horrific that the United States Navy sought ways to prevent it from ever happening again, to prevent human error that would result in the tragic deaths of noncombatants even in a war zone. Thus, they turned to an entirely new weapons system called Cooperative Engagement Capability, an artificially intelligent computer system that, they believed, would eliminate human error once and for all. The computer would do it.

Here's how it worked, this modern incarnation of Tesla's robotically guided missile, futuristic in its elegance, but absent the prohibitions of Isaac Asimov's First Law of Robotics. After all, it was a weapons system designed to be defensive, but designed to kill humans that were attacking it. Because Cooperative Engagement Capability, or CEC, was designed to operate in brown, or shallow inshore, waterways, like the Straits of Hormuz, and possibly in commercial airlanes; no fly-zones notwithstanding, it had to meet the criteria of filtering out land clutter and non-hostile airline traffic within heavily populated civilian environments. It relied on a cross-hatching of radars from different sources lighting up a specific area and a computer matching program to discriminate among hostile, non-hostile, and friendly aircraft. So far so good. Then, once a computer-driven intelligent application determined that a hostile target was in the area, threatening the naval vessels, the computer identified the target, locked on its targeting radar, decided what type of weapon would be deployed and what the launch platform would be, and then issued—absent human intervention—the commands to attack that target. It was supposedly a failsafe mechanism that would not make a mistake.

But it did.

On the night of July 16, 1997, the same type of situation happened again, only this time one of our own planes was shot down. TWA Flight 800, on its way to Paris, had been held up at JFK because its connecting flight from Chicago's O'Hare was delayed due to thunderstorms rolling in over Lake Michigan. As a result, Flight 800 took off an hour late, after Military Operations Area W-105 had gone hot. Absent a warning from Air Traffic Control that W-105 had been turned into a no-fly zone, Flight 800 took off from Kennedy. At approximately the same time, the Army launched a drone, simulating a missile, over the southern coast of Long Island. The cross hatching radars performed according to their programming, locked onto the drone, and transmitted their identification and targeting instructions to the small fleet of naval vessels on station off shore. Among those vessels was a yet-to-be commissioned nuclear-powered attack submarine, the *Seawolf.* The *Seawolf,* submerged, received the CEC data via the Navy's newly developed blue-green water penetrating laser, and the computer designated the sub as the launch platform and initiated the arming of an antiaircraft missile. Then, pinpointing the drone, the computer launched the semi-active radar missile. At the same time the missile was breaking the surface of the water, Flight 800 was climbing through the area. The missile, now airborne and detecting a bigger target in its path and under control of its own radar, deviated from its original course and turned its nose towards the 747 as the crew of the *Seawolf* and the other vessels in the area watched their radar screens in horror. History was repeating itself. Another vessel launched an antimissile missile at the *Seawolf's* missile in a desperate attempt to destroy it before it hit the TWA 747. However, both missiles locked onto Flight 800, one missile hitting where the right wing attached to the fuselage and blowing the number three engine off the wing. Just imagine the panic inside the passenger cabin as the overhead oxygen masks dropped and the frantic pilots tried to right the plane, which had flipped over on its left side once the number three engine was torn off. Then the remnants of the missile's thermite warhead blew through the cabin and ignited the center wing tank. The second missile exploded on the left side of the fuselage, striking the forward section of the aircraft which broke off and tumbled into the sea.

All of this would make for a grand conspiracy theory if not for the fact that the entire event was witnessed by observers on Long Island's south shore and a fisherman on the water that night, who saw the missile rise from beneath the ocean. Forensic evidence obtained by James Sanders showed that there was a streak of thermite residue in the 747 cabin, thermite which could not possibly have been a part of the passenger seat covers. The missile strike on the plane was also witnessed from above by Air National Guard pilots and by the pilot of another commercial airliner.

While what happened in the aftermath of that airline tragedy is beyond the scope of this book, suffice it to say that parts of the FBI and Department of Justice investigation into the causes of the crash are still classified by the government. In a likely effort to protect the highly classified nature of the weapons system that caused the shoot-down, President Bill Clinton was ordered, yes, ordered, to yank the whistleblower protection status from all naval personnel who were on station that night. But the witnesses, the forensic evidence, a video taken by a witness on Long Island's south shore, a snapshot that caught the cruise missile flying over Long Island's Shinecock Bay, and the Navy's own documents paint a picture of what actually happened. All of this only goes to show that 1) the dead past does not bury its dead, as the involvement of the New York office of the FBI in the 2016 presidential election will probably reveal, and that 2) nothing, absolutely nothing, disappears without a trace. For example, a March 20, 2009, article in Space Daily describes the development in Israel of new networked battlefield software system to integrate all situational intelligence sources called SeaCom, seemingly an enhanced modernized version of CEC.

Was the Cooperative Engagement Capability system so highly classified that not even the President or his closest advisors were allowed to talk about it? We know that there are levels of classification of information above Top Secret, levels to which the President is not normally privy. President Clinton once called this the secret "government within the government." Perhaps the Cooperative Engagement system was one of those weapon types. We do know that the crash of TWA Flight 800 was discussed in the White House situation room. What we don't know now is whether the President was specifically advised by military brass not to talk about the crash or to let the NTSB and FBI investigations go forward on their own. We do know, or at least we have been told, that there were such times in the past, specifically when President-elect Jimmy Carter asked then-director of the CIA George H. W. Bush to disclose some top-secret files about anomalous aeronautical phenomena—UFOs—that the CIA might have had in its files. Director Bush refused to answer President-elect Carter's request, telling him that he had "no need to know." Multiple witnesses have confirmed this exchange between Carter and future president Bush. And we remember President Obama's response to ABC late-night host Jimmy Kimmel's question about UFOs. Obama said, "They tell me not to talk about it." Was he joking?

Today, both the US and Russian Navies are developing robotically controlled, artificially intelligent self-guided torpedo-like craft that can stay underwater for prolonged periods, receive updated targeting instructions, seek out enemy submarines, and detonate themselves close enough to destroy them. These are not just on the drawing boards; they are in their testing phases and close to deployment. These weapons systems, although they may seem like a far cry from what Tesla tried to sell to the Navy at the outset of World War I, are actually exactly what Tesla envisioned, self-guided, robotically controlled missiles. However, as Isaac Asimov predicted, the best-laid plans of men and robots "gang aft agley."

Similarly, the Navy is also pre-positioning a system of storing underwater drones in large enclosures on the sea bottom, a type of pre-deployment of a weapons system that can be activated remotely in time of war to seek out and destroy enemy shipping. Building on Tesla's idea of preprogrammed instructions, the drones could be used to intercept enemy submarines, especially guided-missile-launching submarines, to prevent them from reaching their launch points. A plus is that the drones do the job far less expensively in terms of human and hardware assets of engaging enemy submarines underwater.

The United States is so heavily invested in the military potential of artificial intelligence that former Secretary of Defense Ashton Carter repeatedly turned to the high-tech community in Silicon Valley for help in countering cyber threats to the United States. He also turned to them for offensive tactics such as "cyber bombs" to attack an enemy's cyber infrastructure. The latter is important because the theory behind cyber warfare is not just protection but a realization that our only other near-peer superpowers, Russia and China, have developed new generations of weapons that could pose a threat in the near term, a threat that came to fruition in the 2016 presidential election when, according to all of our intelligence services and the FBI, the Russians, under Vladimir Putin's orders, hacked into our electoral process so as to throw the election to their preferred candidate. Looking at these threats from advanced weaponry, the Defense Department has embraced what it calls the "Third Offset Strategy," a reliance on "computer-based high-tech weapons" to offset the technological emergence of China as a near-peer adversary and Russia's re-emergence as a developer of high-tech weapons.[63] The United States, Secretary Carter revealed, is facing enemies with larger inventories of weapons and larger military forces, which means we will have to defend ourselves against superior odds. As a result, our weapons have to be smarter than humans in order to make the correct decisions, either at the battle front or behind the lines.

The development of artificial intelligence poses a unique challenge for the military. Unlike weapons development in the twentieth century, artificial intelligence technology is being developed in the consumer and business sectors. Systems such as Apple's Siri, Microsoft's Cortana, and Amazon's Alexa, while not complete artificial intelligence packages, conform themselves to the most frequent decisions made by users, and will have the ability to predict

---

63 Markoff, John, "The Pentagon Turns to Silicon Valley for Edge in Artificial Intelligence," *New York Times*, (May 11, 2016).

choices their users will make. Because the application of artificial intelligence to the needs of users is taking place outside the military industrial complex so as to address the needs of business and civilian users, Secretary Carter said, "That's different than thirty or forty or fifty years ago when we expected to control the pace of technology."[64]

Lest we think that only the military deploys artificially intelligent systems aggressively, we only have to look closer to home, both at police agencies and commercial enterprises. In modern police work, companies are working hard to provide software to law enforcement that sweeps up information from social media, phone records, and credit-card purchases to build profiles that police can use, often without warrants, to identify suspects and initiate surveillance. But it's not the police who make the decisions about whom to identify, it's an artificially intelligent algorithm that predicts whether an individual is likely to have a criminal intent as measured by his or her actions. This is not science fiction. It is an advanced use of artificial intelligence, a computer only responsible to its internal logic even though it ultimately reports to judicial procedures.

In a Dallas police sniper incident in July 2016, the police, having cornered the sniper in a building, dispatched a bomb-carrying robot to approach the armed and dangerous sniper and detonate itself, killing the sniper. Although Dallas Police Chief David Brown told reporters that his negotiators had made repeated attempts to talk the sniper into surrendering, he felt he had no choice but to dispatch the robot. But, as reported in *Time* magazine, there were a number of ethical questions that were raised, not the least of which

---

64 Ibid.

was that this entire robot detonation incident in a civilian police engagement was nothing less than an evolutionary event. A robot was deliberately dispatched to kill a human being. Is this just a real-life version of *RoboCop* or are we looking at something far worse, the beginning of something we can't stop?[65]

What this discussion shows is that even as we look at the works of Tesla and Edison, their theories about how things work and their predictions about the nature of life and how life can be imitated by machines still resonate today and affect the world we live in. Artificially intelligent systems have brought us into the future. But they are the future based upon what both Edison and Tesla envisioned, as both scientists foresaw the rise of machines and their amalgamation with human systems.

EDISON'S PREDICTIONS FOR THE FUTURE

As early as 1911, Edison predicted in both *Cosmopolitan* magazine and the *Miami Metropolis* that steam-powered railroad trains would one day be powered by electricity and would travel at fantastic speeds. His prediction will finally be realized in the US, according to Elon Musk, who, among other innovators of technology, has laid out plans for high-speed magnetically levitated trains running between Los Angeles and San Francisco as well as along the Northeast corridor from Boston to Washington, DC.

Edison predicted that the printed book would be replaced by electronic books, which might be how you're reading this, delivered almost instantly from publishers to buyers and readers. Travelers, he predicted, would be able to reach their destinations by air, flying

---

65 Edwards, Hayley Sweatband, "When Can Police Use a 'Bomb Robot' to Kill a Suspect?" *Time*, (July 8, 2016).

at speeds as fast as two hundred miles an hour. (We know that modern military aircraft have already exceeded the speed of Mach 3 and that commercial airliners fly at speeds in excess of 500 knots.)

Steel, he said, would replace most wood in furniture. In fact, he predicted that infants would be rocked in steel cradles and that furniture itself would be so light that one person could easily move it from place to place. One only has to imagine the assembly of lightweight build-it-yourself furniture such as one might find at stores like IKEA to see that Edison was right on target in this prediction. Similarly, Edison predicted that steel-reinforced concrete would be the construction material of choice for new buildings, replacing brick and wood frame. This, too, has come to pass. Edison predicted that telephonic technology would usher in a new age of information sharing, turning the device that Bell invented and that Edison improved into a type of knowledge dispensation device. With such a device, callers might be identified not by their names but by other forms—think of today's avatars in a telephone-based Internet universe—and that one would use telephones to get price quotations from various markets, especially the stock and gold markets.

Gold, Edison also predicted, would cease to become a precious metal because the twenty-first century's rediscovery of the protocols of alchemy would mean that scientists would solve the age-old mystery of transmutation from lead into gold. Instead, according to the late Terrence McKenna, it was alchemy itself that changed, morphing from the transmutation of metals into the morphing of science itself to recreate what science perceived as reality. By the 1940s, Nazi engineers in Poland, for example, were trying to fabricate the transmutation of heavy metal in liquid form into radioactive elements to build the fuel for an atomic bomb. The process, which took place inside a hollowed-out mountain in the Owl Mountains in Poland, was only stopped as Soviet troops approached and SS Colonel Kurt Debus, the manager of the Nazi project, called "Die Glocke," fled along with his research to the allies as part of Operation Paperclip. If this nuclear enrichment device had worked in time, it would have provided the Nazis with fissionable material they could have used, theoretically, at least, to build a nuclear weapon.

Machines, Edison told *Cosmopolitan*, would assemble products themselves, robots on assembly lines rather than simply fabricating separate individual parts for human workers to assemble. In this, Edison envisioned the robotic factory, the staple of modern car manufacturers across the globe. Edison's vision of the automated factory, machines building other machines, was at least a decade ahead of Fritz Lang's dystopian silent film *Metropolis*, in which human beings fed the machines that powered the society that kept the worker humans as subjects to the demands of technology.

Edison didn't confine his prognostications about the future of technology and automation to the world of business, industry, and consumer product manufacturing. In 1911, he also imagined a world in which many socially divisive issues would be resolved. For example, he predicted that technology would eliminate the need for war and that the twenty-first century would see a time of peace. Of course, World Wars I and II had not taken place yet and the development of nuclear weapons had not threatened civilization. But Edison predicted that after a series of climactic battles, the world would settle down to peace because that was the only alternative to destruction. And, finally, technology itself would be the means to end poverty in civilization. Human beings, he said, would enter a period of plenty in which the class differences separating haves from have-nots would be eliminated.

This looked to be a rosy future, indeed.

CHAPTER 13

Mediums, Minds, and Machines in the Roaring Twenties

By 1920, as Thomas Edison entered the last decade of his life, he was at the forefront of the triumph of science over the age of pure spiritualism, even though the two streams of thought were coming together.

The twenties were the decade during which the world, in which Edison had created

TIMELINE

June 28, 1919: Treaty of Versailles ends World War I
October 28, 1919: National Prohibition Act passed
January 16, 1920: Prohibition begins

his greatest inventions and launched the three great industries that would define the twentieth century, had changed dramatically. In the aftermath of the Great War, World War I, there was a tremendous sense of malaise in the West.

World War I was the first war involving significant loss of civilian population. It was the first mechanized war and the first war where a weapon of mass destruction—poison gas—was used. World War

I was the first official air war that involved not just chivalrous pilots saluting one another from their cockpits as they engaged in dog-fights, but a war in which bombs were dropped on noncombatants who had no means of escape. World War I was a war of machines, of tanks and motor vehicles and huge pieces of artillery drawn by train to wreak havoc on cities. And it was a bitter war fought in the trenches of France and Belgium, in which waves of infantry expended themselves in hopeless charges, winding up stretched across barbed wire and lying in pools of mud inside the craters created by the explosions of artillery shells. This was a war without glory, without chivalry or honor, and whose heroes would return home with lungs rotting away from mustard gas or shell-shocked into a near somnambulistic state. We call that post-traumatic stress disorder today, and it influenced the entire postwar generation.

It was no wonder that the generation of youth that fought that war called itself the "lost generation." Just imagine an entire generation desperately suffering emotionally from the aftermath of a new kind of war, from deprivation, and from the transition, brought about by war, from an agricultural country to a mechanized, urban one. It was as if all the elements of a collective post-traumatic reaction came together.

PROHIBITION AND EXCESS

The 1920s was also the decade of Prohibition in America, when alcohol manufacture and distribution were illegal and the market for providing alcohol was left to organized crime. Thus, much like today, the illegality of a substance actually created a huge contraband market, with organized crime cartels forming their own local economies and taking over whole cities through bribery and corruption. That alternate economy run by the mob also created a great sense of distrust in the government and contributed to the malaise of the 1920s; this, despite the growth of Edison's new motion-picture industry, which, by the end of the decade, would also be penetrated by the crime cartels from Chicago and New York seeking to organize the unions.

The 1920s was a decade of financial excess. Although this may have seemed like an anathema to a conservative businessman like Edison, it might also have seemed like an opportunity to the entrepreneurial side of Edison. His company, General Electric, would have benefited from the free flow of investment dollars in the publicly traded stock market because the margin calls on stock purchases were incredibly favorable to buyers. For example, for an investment of just 10 percent, ten cents on the dollar, one could purchase stock on margin, betting that the surging prices of stocks would ultimately make up for the ninety percent loan from the brokerage house. Stated another way, if you bought stock at $1,000 at 10 percent margin, owing 90 percent of the value of that stock, and then the price of that stock increased so that your $1,000 value became $2,000, you could pay back the $900 you borrowed and make a hefty $1,000 on the deal when you sold the stock. Multiply this and one can see why traders invested heavily on favorable margins. But, just like the financial housing bubble that burst into the Great Recession of 2008 and the leveraged buy-out crisis of the 1980s, all good things come to end when the bank calls in the loan. This happened in 1929, thus initiating the stock market crash.

But during the Roaring Twenties, Edison was a witness to this perceived great affluence and, perhaps, knowing about previous stock market crises, he was also wary of financial overindulgence. He consequently sought an invention that might not depend upon heavy investment. His use of a simple light-beam projector and a photoelectric cell, as primitive as that sounds, would have been just the type of apparatus, had it worked in the design phase, to open up an entirely new market, at a price that individuals could afford. It would have been just like any home appliance, albeit to contact the spirit world.

THE THIRD CULTURE

As the scientific industrial revolution—as opposed to the scientific intellectual revolution of the sixteenth century—that began at the end of the nineteenth century took hold, it did not entirely sweep away the trailing edge of the Great Age of Spiritualism. In fact, the two intellectual and cultural movements actually merged by the 1920s, creating a third culture, a neospiritual movement that encompassed the science of the unseen—atoms, quantum theory, the movement of electrons—with the spiritual and the world of noncorporeal entities. It was only in 1935, a few years after Edison's death, that the theory of quantum entanglement was first proposed. However, it seems that Edison might have already had an inkling concerning the nature of quantum entanglement, the relationship between separate quanta of matter that behaved with such a synchronicity that if one quantum spins one way, then its partner, or partners, would perform in exactly the same way. And if quanta could be entangled across space, why couldn't they be entangled across time? This was a theoretical underpinning to Heisenberg's theory of uncertainty as well as a scientific rationale for the phenomena of remote viewing as articulated by one of its early developers, Ingo Swann.[66] It was called "spooky action at a distance" or cohesion, and it became the basis for the design principle behind Edison's spirit phone.

---

66 Swann, Ingo, *Penetration*, (New York: Ingo Swann Books, 1998).

EDISON'S THEORIES OF DESIGN
FOR THE SPIRIT PHONE

Edison had to bring a number of theories together to engineer his spirit phone. First, he had to refine his thoughts about the cohesion of subatomic bits of matter, things he could not see but believed existed. He formulated an idea that if human consciousness, sapience, and the essence of personality were the result of bits of matter coming together, a purely materialistic theory, then why would the death of the body destroy the cohesion among the particles? Perhaps, even for a short time after death, the bits of matter that made up a human being's consciousness still stayed together, which is what the UK's Sir Roger Penrose has theorized. If they stayed together, then how would we know they stayed together?[67]

Edison's second theory: What if these bits of matter were charged electrically? They might simply be electron-based. If electrons or similar units, then they would certainly give off an electrical charge, a negative charge.

Third theory: What would happen if a cohesive bundle of negatively charged particles crossed through a tightly focused field of photons, light? If they did, they should generate an interference with the photon stream and register a form of electric charge on a receiving cell or meter. If all that were to be true, then an ideal experiment would involve building a device that generated a narrowly focused photon field toward another object, at the end of which was a cell to register anything crossing through that field. Then, if there were some sort of meter attached to the cell, Edison

---

67 Penrose, Sir Roger, *How Consciousness Became the Universe: Quantum Physics, Cosmology, Neuroscience, Parallel Universes*, (London: Science Publishers, 2017).

should get registration on that meter, something he could measure and prove to exist scientifically so as to be able to repeat it.

This seemed simple enough, but there was still one question that had to be answered before anything else: how would one know whether an entity, a cohesion of matter, was actually in the area so as to prove the theory of the spirit machine, a proof of concept? Edison would have to find a way to summon the cohesion of matter, a spirit of the departed. And this is where his science merged with the supernatural. How do you summon a spirit?

THE ROLE OF MEDIUMS AND CLAIRVOYANTS

Edison believed that mediums who claimed they could communicate with the dead were charlatans. And although he had heard the stories of the Fox sisters in the latter nineteenth century and had seen America's fascination with communicating with the dead via mediumship through the early twentieth century, he did not believe those stories were true. He thought that mediums, particularly the Fox sisters, who had admitted to faking their claims, were simply nightclub performers. However, if Edison could test the efficacy of using mediums to summon departed spirits, then he could use their efforts to test out his two theories: 1) that bits of electrically charged matter comprising the consciousness of the departed still remained cohesive after death, and 2) that if this cohesion of electrons passed through a photon field they would register on that field. If these two theories were true, then might the meter he planned to put at one end of the field attached to a photoelectric cell be able to register that presence of a cohesion of electrons?

Even if he could prove all of the theories by developing a machine that would register the presence of cohesion of electronically charged particles, would that be enough? How could you differentiate a random spike on an electrical meter with the scientific proof that the spike indicated the consciousness of a departed spirit? In order to do that, you needed a means of communicating with that spirit via another channel so as to confirm the results. Moreover, what would the results be? How do you actually speak to a departed spirit through a device? A spirit guide can at least translate any telepathic messages, but a telephone? An entity actually has to speak into one end in order for sound to come out a speaker. Can departed spirits vocalize?

Edison figured that if his machine worked, he could devise a code, maybe even starting with what the Fox sisters did, a yes/no dialogue in which two registrations would mean yes and one registration would mean no. From there, perhaps he could move on to a Morse code to develop an actual language. And, thus, by stages, Edison believed that he could figure out how to build a machine, based on scientific principles of conductivity and photoelectricity and grounded in the theory that matter is never destroyed or created but only changes form, that would communicate with the dead.

EDISON'S KILLER APP

His next goal was to figure out what the market might say about such a device. If the market didn't exist, then he would make a product that would create the market it sought to satisfy. Stated simply, Edison invented what today we call the concept of the "killer app." Edison invented the motion-picture camera and from that created the motion-picture entertainment industry. He figured out how sound waves could be translated into physical grooves on a rotating foil cylinder and thus created the recording industry. Now, in the wake of World War I and the horrendous loss of life, Edison saw another market, the market of the bereaved.

Hence, combining theories of Spiritualism and Materialism, set within the context of a worldwide malaise and grief over the loss of life and a way of life after World War I, Edison commenced his development of the spirit phone. In order to do that, however, he had to determine that there was a real science, an absolute materialism underlying the practice of spiritualism.

Was there really such a science and where is it today?

CHAPTER 14

The Science of Mediumship, Clairvoyance, and Remote Viewing

For Edison, a man of science whose whole career of inventions was premised on the discernible and quantifiable flow of electrons, it would take a great leap to impart a basis in science to spiritualism and the paranormal. Yet, by the early twentieth century a considerable body of scientific research and even empirical evidence had built up promising just such a scientific premise to the paranormal. And this is what was necessary for Edison to come to grips with the underlying theory of the spirit phone.

RESEARCHERS INTO THE
SCIENCE OF SPIRITUALISM

1794: Emanuel Swedenborg enters his spiritual phase
1850s: Sir William Crookes experiments with spectral spiritual analysis
1880s: American Society for Psychical Research
1884: Sir Oliver Lodge experiments with psychic transference
1915: Bert Reese tested in court
1916: Sherlock Holmes author Sir Arthur Conan Doyle embraces spiritualism
1917: Bert Reese becomes an assistant to Edison
1920s: Harry Houdini becomes a spiritualist debunker

1920: Edgar Cayce treats President Woodrow Wilson

1934: J. B. Rhine publishes *Extrasensory Perception*

1980s: Remote-viewing program at Stanford Research Institute

1981: Medium George Anderson provides grief counseling

A common belief is that mediumship and channeling are either completely paranormal events not subject to science, or they're complete hoaxes. But neither of these two beliefs is necessarily true. Even some true believers assert that mediumship and clairvoyance are not subject to the rules of science but lie outside the realm of materialistic proof. We know, for example, that Edison, though he called mediums "charlatans" in his diary, had employed the services of medium Bert Reese to communicate with the unseen spirit world. Thus, if a scientist and electrical engineer and inventor such as Edison dabbled with mediums, doesn't that imply that he saw some science to the paranormal?

AMERICAN SOCIETY FOR PSYCHICAL RESEARCH

Although there is a concerted scientific approach to studying the paranormal today, specifically by Dean Radin,[68] we also know that mediumship during Edison's lifetime had already been studied scientifically. Much of this was done at the American Society for Psychical Research (ASPR) in the 1880s, where researchers devised experiments to rule out chicanery. Much like the way modern skeptics try to eliminate the possibility that self-described clairvoyants can use clues from their subjects to reach answers, these researchers sought to control the conditions under which those being tested arrived at their conclusions. Thus, researchers

---

68 Radin, Dean, *The Conscious Universe*, rpt. (New York: HarperOne, 2009).

sought to evaluate their subjects' results objectively and statistically. This is the same type of research being conducted today, especially in light of the CIA-funded Remote Viewing Program at the Stanford Research Institute back in the 1980s, as described by retired Army major Paul H. Smith, PhD, in his book *Reading the Enemy's Mind.*[69]

Among the questions raised during the scientific evaluation of mediumship and clairvoyance was whether communicating with the dead was really proof of life after death or simple mental telepathy? Was the clairvoyant actually picking up messages from the other side or was the clairvoyant picking up thoughts from a subject's mind about the departed person? In Edgar Cayce's transactions with his clients and subjects, including President Woodrow Wilson after Wilson had suffered a debilitating stroke, it seemed clear that he was picking up impressions from his subject. Was that what was happening with other clairvoyants? This is the science that Edison sought to determine with his spirit phone.

Though the ASPR claimed that their research showed mediumship was real and could be proven, their research did not persuade skeptics, who argued that the results were skewed and that there were no real scientific controls. Edison was among the skeptics, because he believed in the basic premise of scientific testing to prove a theory: repeatability. If the same exact result couldn't be reached by repeated attempts, then the results would be false and nonscientific. Hence, for proof, there had to be both statistical reliability, which meant that the results could not be simply random but more than a 50/50 success rate, and the success rate, if any, had to be repeatable objectively by others.

69 Smith, Paul H. *Reading the Enemy's Mind*, (New York: Forge Books, 2005).

J.B. RHINE

The next researchers of note were J. B. Rhine and his wife Louisa, both with doctorates, who designed and ran an innovative laboratory at Duke University that promoted scientific study of extrasensory perception (ESP) during the 1920s. Their research, published by the Boston Society for Psychic Research, quickly became a topic discussed in popular publications. The Rhines tested their respondents with what they called "Zener cards," which had different shapes on them, such as a wavy line, a cross, a circle, a square, or a star. Subjects could not see the shapes on the card, but had to determine them using what the Rhines believed were psychic abilities or ESP. Theoretically, subjects guessing by chance would have a one in five chance of guessing the right shape, because there were five different shapes. They used shapes because they felt that the standard designs on the cards should be completely neutral so as not to arouse the emotions of the subjects. If the subjects had a high percentage of accuracy in determining the hidden shapes on the card, according to the Rhines, it would mean that, statistically, there was something else at work besides pure chance—a higher ESP capability, in other words. Although the Rhines' results were subjected to critical analyses from statisticians who argued that the methodology was faulty, other experiments using similar types of shape identification also showed that the success rate was greater than random chance. Hence, there was some science, albeit rebuttable, to what the Rhines were doing.

Important for our purposes here is that these experiments were taking place at the time when Edison was perfecting his spirit phone. Moreover, because the Rhines' experiments were being picked up by the popular press, Edison certainly would have known about them even if he had not read their original report. We also know that the entire question of ESP fascinated Edison.

SIR WILLIAM CROOKES

One individual with whom Edison did correspond was the British inventor Sir William Crookes, a scientist—chemistry and physics—who was a highly regarded scholar and a member of the Royal Society of Chemistry. His specialty was spectroscopy, a dissection and analysis of an element's light emanation. The spectral analysis of an object can determine what it's made of. Crookes applied spectroscopy to the paranormal by analyzing the light from mediums, similar to an aura, which allowed him, he said, to determine the material of a spirit as well as that of the medium who said he was channeling the spirit. Thus, Crookes claimed he was able to see changes in the matter around him during spiritual encounters. His paranormal experiments claimed to have found evidence of what he called "spirit photographs."

Whatever Edison might have gleaned from Crookes most likely bolstered his theory for the spirit phone. If Crookes's scientific approach to the presence of a spirit around a medium was to identify specific material based on a light pattern, then might a projected beam of light offer the means not only to identify the presence of a spirit but also to record that presence on a meter wired to a photoelectric cell? This was the challenge. Thus, we can assume that Edison did not come up with his idea for a spirit phone in a vacuum. Rather, he had evidence of its possibility based upon Crookes' own spirit photographs, real science based upon the analysis of the dispersion of light. Again, this was real science applied to the validation of the existence of the spirit world. The question is, what would Edison do with the qualitative analysis that Crookes's work produced?

Though older, his mind remained sharp. His rivalry with Tesla continued to drive him, even though Tesla was outpacing him in venturing into unknown scientific territory, particularly with theories that radio waves are eternal and carry messages from extraterrestrials as well as from spirits. Could Edison, now a corporate giant, simply accept Tesla's presence and retire with his memories of great accomplishments? Not a chance. That was not his manner nor personality. There was more to do, and thus he announced his next proposed invention, causing a "national sensation." Edison's announcement generated nationwide sentiment, both among families, who had lost loved ones and sought Edison's help in contacting them, and among skeptics and religious fundamentalists, who wrote Edison accusing him of witchcraft. Asked if he believed in life after death, Edison was quick to answer that he was a man of science, and that his idea for a device to communicate with the dead had nothing to do with belief.

Skeptics were especially critical about Edison's pronouncement because it flew in the faces of those who resisted talk of an afterlife or spirit world. Powerful opposition came from conservative organized religion, which held that heaven and the afterlife belonged in the domain of theology. Atheists, of course, debunked spiritualism outright. But for Thomas Edison, looking at the writings of Swedenborg and Blavatsky and reading the work of the Rhines and Crookes, spiritualism was a form of science even if it did not measure the passing of electrons through a circuit. But what if it could?

Edison, as noted earlier, had a curiosity about what happens to us after death. Is there an afterlife, a spirit world? Might there be a heaven and a hell? Can those of us who are alive communicate with the deceased? At one point Edison had even attended séances conducted by Bert Reese, who worked in his laboratory. It was a fascinating idea to bring science to a study of the afterlife. And it would be even more fascinating, Edison thought, to build upon the work of Sir William Crookes, to prove with a light beam and electrical current what Crookes claimed he had proven with spectral analysis. Edison was following in a great tradition.

SIR OLIVER LODGE

The year 1920 ushered in an interesting era in a changing America. Among other things, there was a rebirth of interest in spiritualism. One of the most famous scientists of the time, Sir Oliver Lodge, had lost his son Raymond during World War I and sought mediums in the hope of communicating with him. Lodge was particularly focused on the "survival of memory," something that must have inspired and might have been a contributing influence to the modern theories of life after death promulgated by Sir Roger Penrose almost a century later.

New ideas such as atomic energy and quantum physics had come to the fore, and with them new names, including Einstein, Freud, Jung, and Max Planck. Edison, ever the competitor, knew he needed something new, something spectacular. The spirit phone, he told several science magazines, would be built based on scientific tenets. He fully expected it to work.

GEORGE ANDERSON

In 1981, fifty years after Edison's death, a New York man in his late twenties, George Anderson, agreed to a series of tests to determine if his ability as a medium was genuine. Over the next several years Anderson underwent dozens of experiments, many conducted live on radio and TV. The tests covered a wide range of Anderson's psychic gifts, and were supervised or observed by an array of scientists, physicians, psychologists, various medical personnel, engineers, skeptics and debunkers, magicians, parapsychologists, and members of the clergy. Some of the experiments utilized scientific technology such as EEGs, MRIs, and CAT scans. The goal was to determine if there were changes in brain activity both before and during times the medium claimed he was receiving messages from the departed. To the surprise of those involved in the testing, the medium had demonstrated a high degree of accuracy in his communications, and displayed brain state changes. In this way, not only did he beat the odds of random chance, he demonstrated medically as well as empirically that there was a science to the paranormal.

In several other medical and scientific settings in the US, Canada, and Europe, others were also performing experiments about paranormal abilities, and such related phenomena as near-death experiences. Too often, however, successful or positive results received scant public attention or went unreported. Despite this, mediumship and other psychic events are slowly being accepted by an increasing number of scientists; more people are open about their experiences; and the paranormal, especially the question of life after death, has become a popular topic in the media.

Many think that the testing of George Anderson marked the beginning of a modern age of mediumship. Thanks to advanced technology as well as more scientists having open minds about the possibility of communication with an afterlife, types of testing that had not been possible in the nineteenth century could now take place.

The first time Anderson demonstrated his mediumistic ability for a group of friends and witnesses, he was uncannily accurate. How did he know of the habits of a subject's late wife? What gave him a clue that another subject had been abused as a child? How did he determine that another subject had been a twin, whose twin sibling died very young? Did some spirit entity tell him that a subject's uncle was dying of prostate cancer? Was it a trick? Could the medium somehow have had advanced knowledge of the subject he was "reading"? Was he picking up verbal or visual cues from a subject? These were all reasonable questions. Precautions had been taken in anticipation of such objections, however. Anderson and the subjects did not know each other, which ruled out the chance that the medium had gotten any information ahead of time. And in the case of readings over the telephone, the only answers a subject could give were 'yes' or 'no.' Thus, there was no chance of visual or verbal cues tipping off the medium.

Since little was written in the 1980s about the science of mediumship or testing someone for mediumistic ability, it was necessary to create tests with the help of everyone, from scientists to skeptics. The tests that were organized sometimes seemed deceptively simple. Objections from skeptics and debunkers were always a factor to contend with. For that reason skeptics were included in the group of experts, including one who was both a psychic debunker and magician with the skill to recognize any deception. The goal was not to prove that the medium was genuine, it was to demonstrate whether the medium was behaving in a fraudulent manner or whether there might be another explanation for the medium's successful results. Every precaution against chicanery was taken.

Anderson was a frequent radio guest, talking about his abilities and inviting callers to ask him questions. This was the beginning of late-night open-mike radio. One night, while answering listener questions on air at a Long Island, New York, radio show, Anderson received a call from an anonymous male. Anyone who phoned in to the live shows was asked to say hello or say some innocuous phrase so as not to give the medium any name, advance information, or even the hint of a foreign accent. The first word Anderson heard from the caller was, "Hello." The medium's answer was to suggest that the caller had an unusual amount of psychic ability. The conversation between the two continued for a few more moments before the person who phoned in identified himself. He was Robert Petro, one of the nation's most gifted and successful psychic-clairvoyants. How did Anderson know that the anonymous caller was someone with a high degree of extrasensory perception?

In another test, the goal was to determine if distance was any factor in the medium's ability or accuracy. Most of those phoning in were from the greater New York area. This time, however, the female voice at the other end of the phone line was calling from Hawaii. The arrangement was made by the show's producer without the knowledge of either the host or Anderson. The medium's accuracy was no different than if the caller had been around the corner. Clearly, distance did not matter.

Following each experiment, typically within the next few days, the group of scientists, skeptics, and parapsychologists gathered to analyze and theorize the test's outcome. Members of the group often disagreed; but that was the purpose of discussing the medium's conclusions from several different disciplines, including theology.

One day Anderson was visited by a thin elderly man who was dressed in shabby clothing that reeked of alcohol. His sparse grey hair was unkempt and he was unshaven. Anderson was at first startled by the stranger. This was not the kind of person who typically sought Anderson's psychic readings. The medium stared at his strange visitor. Was he homeless? Was he a drunkard? Within moments, Anderson asked the man whether there was a reason he saw the spirit of Sigmund Freud behind him. Perhaps the visitor was a mental patient? The stranger acknowledged that yes, psychically seeing Sigmund Freud made sense. Finally the man confessed. He'd heard George Anderson on radio and was convinced his medium-ship was trickery or outright fraud. The man admitted he was nei-ther homeless nor a drunk. The stranger had disguised himself in hopes of testing or fooling Anderson into picking up on incorrect visual clues. In fact, he was a respected and successful psychiatrist. That explained why Anderson clairvoyantly saw what he presumed was the spirit of the late famed psychiatrist Dr. Sigmund Freud. The doctor was both stunned and impressed. He had not duped the medium by pretending to be someone else.

One evening, also at the radio station, Anderson was intro-duced to a handsome, middle-aged man meticulously dressed in a three-piece suit, his hair perfectly groomed. He was introduced as "Joseph," who was visiting the United States from London, where he was in banking. Anderson stared at the visitor for a few moments. It should be noted they'd never met before, and that Joseph did not believe in mediumship, an afterlife, or a spirit world. Anderson suddenly asked, "Why do I see John Lennon's spirit near you?" "I don't know," Joseph answered. Anderson insisted, then went on to explain that the ex-Beatle's spirit was psychically com-municating an account about how he and Joseph were boyhood friends in Liverpool. Psychically, Anderson heard Lennon tell of one incident when they were at a party and Lennon threw ice cubes down Joseph's back as a joke. Joseph recalled the incident well, but could not understand how Anderson could possibly know it. The medium continued by recounting several personal incidents, includ-ing a family death years ago. By now Joseph was shaken and pale. Who was Joseph? He was not a London banker at all. He worked in marketing and promotion in the music and entertainment indus-try, although he did not look or sound it. Joseph was certain the medium could not have known about the ice cube incident. It was another mystery in the science of mediumship.

It was not unusual for Anderson to receive phone calls and letters throughout the 1980s and 1990s from scientists, atheists, debunkers, and skeptics. Typically they were individuals who were certain Anderson, and other psychics and mediums, were all frauds. Anderson and his radio host developed an answer for them. Since most of the debunker/skeptical callers had access to technology, they were asked whether they wanted to test any of the mediums, including Anderson, with their equipment and announce their findings on live radio or TV.

One phone call from an angry doctor threatened that if Anderson were a fraud, the doctor would take legal action. As a result, Anderson underwent a series of tests that employed thermography, which involves a device that tests heat emanating from the body and color codes hot and cold areas. Anderson was instructed by the doctor in charge of the testing to give psychic readings to each of several subjects known only to the physician. Each subject had a specific health problem which Anderson was to identify; heat output from a part of Anderson's body would register on the thermography machine's screen, indicating that the subject had an issue in the same area. The subjects were not permitted to speak or acknowledge any answer Anderson gave. Record-keeping was maintained in writing and on video. Anderson was dressed only in a pair of shorts and was sitting in a chilly, temperature-controlled room. He could not see the thermography screen or the results. In other words, not only was he deprived of any physical feedback to his questions, he was deliberately kept chilled so that his body heat, if it registered, would show up dramatically on the thermograph.

The first subject was a young woman. As soon as Anderson saw her, his left breast heated upon the thermography screen, indicating that he was experiencing neuromuscular thermal reactions to sensations he was receiving from the test subjects. Anderson correctly identified each subject's area of illness until the last subject, a middle-aged woman. On the screen one could see Anderson's throat heat up as he told the woman he psychically sensed that she had a problem with her throat and back. The physician marked that as wrong. She was the physician's own patient, but not for any difficulty in the throat area. Still, Anderson's results were statistically impressive. But wait. The session was not over because the woman wanted to say something. Not known to anyone present, shortly before she entered the room with Anderson she became nervous and a lozenge she'd been sucking on became stuck in her throat. When Anderson's throat heated up, as indicated by the color red on the screen, he psychically experienced momentary pain in his throat. This was unknown to the physician, but known to the woman and correctly picked by Anderson. The physician in charge of the testing was left speechless.

At another time and location, under the supervision of a physician and technician, Anderson was again tested with the thermography device. Once more he was accurate with anonymous subjects. In one experiment he was tested against a skilled magician who was unable to register more than minuscule temperature changes. It was also valuable to test the medium several times in controlled settings, to determine if there was repeatability in his psychic demonstrations.

A neurologist at a New York hospital was skeptical about Anderson's purported ability to contact the afterlife. But, after thousands of psychic readings, what accounted for the medium's high degree of accuracy? This doctor was curious about what changes—if any—had occurred in Anderson's brain during the time he said he was receiving messages from the departed.

It was not a trivial question. Mediums and clairvoyants often claim they must be placed in a trance before they can psychically communicate. Skeptics argue that the trance state is simply an affectation or false display to impress the public. Anderson never appeared to enter a trance state prior to his readings. Were their changes in his brain when he claimed he was hearing or sensing spirits?

Anderson was connected by electrodes from his head to an EEG, a machine that displays brain wave states at various times. For example, our brain waves are more active (Beta or Alpha) when we are awake or active, but when we are relaxed (Theta) the brain waves slow, and when we are in a deep sleep (Delta) they are at their slowest.

Anderson was accurate with the subject he was reading. What interested the medical team as much as the accuracy of his answers was the fact that although the medium remained awake, registering Alpha waves, his brain wave state slowed to Theta when he said he was communicating with a departed relative of the subject, a young man chosen by the neurologist.

Other tests subsequently revealed that the mediumistic ability was located in the right side of the brain, known as the right temporal lobe. One series of tests conducted on several mediums by a clinical psychologist found that many of them receive messages from the departed in the form of symbols, much the way we dream. The right side of the brain is recognized as the lobe that deals with dreams, artistic, and creative processes. It may also be the side of the brain that operates when we pray or have religious or mystical experiences, which is why some scientists have identified the right temporal lobe as the "God spot." It appears that the so-called "right brain" is where psychic or paranormal messages are received or sent, which makes perfect sense according to psychologist Julian Jaynes, who argued that it was messaging from the right hemisphere of the brain that pre-linguistic humans believed were impressions received from their deities.

One series of tests involved a machine called a "random events generator." Anderson's task was to tell the computer operator what event and date the machine was spinning out in another room out of sight of the medium. Anderson was unable to answer accurately when, for example, the random events generator indicated "July 4, 1776, Declaration of Independence." Anderson replied, "1941." That was wrong, of course, since December 7, 1941, referred to the attack on Pearl Harbor. Strangely the next item to appear was December 7, 1941. And thus the exercise continued, until the computer scientist realized that the medium was answering in the future. So, for example, if the machine generated "1865...President Lincoln assassinated," Anderson would answer 1492." The next item would be "Columbus...1492." What had occurred? For whatever reason, the medium was moving into the future by predicting what the machine would generate. As strange as this seems, it was verified. This test also foreshadowed what Army remote viewer Paul H. Smith would experience in May 1987, when, during a training session, he remote-viewed an event that actually took place days after he viewed it. He next remote-viewed an event that took place in 2016: the close orbiting of Saturn by the Cassini robotic space probe. Did George Anderson and Paul H. Smith remote view into the future so as to gain impressions in the present? Was this just another form of psychic quantum entanglement over time rather than space, a variation of Einstein's theory, or was it a version of Werner Heisenberg's uncertainty theory, in which the future controls the past, also known as retrocausality? Whatever it was, it goes a long way to suggest that Edison was correct in using quantum theory as a basis for perceiving disembodied entities, just as Penrose is today.

One afternoon, Anderson was being driven by an engineer to a house where he was to undergo yet another round of tests. A couple of blocks from the location, the engineer thought it would be interesting to find out if Anderson had sufficient clairvoyance to see into the house even though they were still a block away. The white split-level suburban home had its windows covered by drapes. There was no way to see inside.

"Can you psychically describe the interior of the place we're going to?" the engineer asked, pointing to the house as it came into view.

Anderson stared at the white-painted building, then said he'd never seen anything like what was inside. He explained that all he saw were "dolls." The entire house was covered with what looked like thousands of children's dolls. It seemed like an unusual answer.

Once the engineer and the medium entered the house, the answer became apparent. The house was, indeed, covered with dolls. They were on shelves, counters, table tops, sofas, anywhere they could be fit or safely placed. The dolls represented one of the largest private collections in Long Island. Once again, Anderson was correct.

One night on the radio show a math analyst sat in to determine how accurate Anderson was statistically. The young man employed a computer and strict rules to learn if the medium could be guessing when he gave callers information such as names, relationships to deceased loved ones, causes of their passing, even occupations. At the end of the session, the analyst concluded that there was only a one in more than 4,400 chance that Anderson could be guessing.

There were other tests and experiments that found unearthly sounds on tapes made during efforts to communicate with the afterlife. One group of experiments employed infrared photography that discovered auras around several mediums while they were engaged in readings.

REMOTE VIEWING

The study of the science of the paranormal, whether called mediumship or clairvoyance, continued through the twentieth century. In fact, such studies were even funded by the United States intelligence services and by the military. In the 1980s, researchers at the Stanford Research Institute tested celebrated psychic Ingo Swann, who had approached them with an idea to use what he argued were the psychic gifts all human beings possessed to view distant locations and engage in a form of psychic surveillance, which is what some inside the US intelligence community believed the Russians were already doing. This process was called "controlled" or "coordinate" remote viewing, in which a subject was not told what coordinates were in the mind of his or her partner, but simply to open up one's mind to what impressions flowed in. In other words, treat the mind as a kind of tuner to pick up a signal. Where the signal came from or what it was, was not the purpose of the experiment. It was only to record the impressions as they flowed in.

Ingo Swann was so successful at this that the researchers were more than impressed. In fact, they tested him by seeing whether he could read or control the readings of shielded instruments. When he was able to accomplish this, despite the fact that he could not see the instruments, the researchers were even more impressed—so impressed, in fact, that they applied for funding from the CIA, using the scientific theory of the "bubble universe" and perhaps string theory and quantum entanglement to explain how the process worked. The CIA was fascinated enough to fund the research. Eventually, a remote-viewing team, now called coordinate remote viewing, because the trainer was only equipped with a set of coordinates on a slip of paper and nothing else, was assembled at Fort Meade by the United States Army. According to retired Army major Paul H. Smith, who heads up his own remote-viewing academy, called Remote Viewing Instructional Services, they met with a success rate beyond the level of chance.[70] In the twenty-first century, remote viewing is a tool of law enforcement agencies, corporate entities, and the subject of books and movies. It has become popularized.

The nexus of retrocausality and remote viewing (neither term was in use at the time) was tested back in the 1960s with children who were given a single marshmallow and told that if they could wait fifteen minutes before eating it, they would get another marshmallow. But if they ate the single marshmallow before then, that's all they would get. While the purpose of the experiment was to test a child's ability to assert self-control, to defer a reward for a greater reward in the future, scientists further suggested that the experiment also indicated a child's potential to show empathy for his or her future self, neurologically related to self-control on a brain/bio level.[71] However, in terms of Heisenberg's theories of

---

70 Ibid. and Smith, Paul H., "Remote Viewing, Possible CE-III Event," *UFO Magazine*, 20 (5) (December–January, 2006), 48–55.

71 Young, Ed, "Self-Control Is Just Empathy with Your Future Self," *The Atlantic*,

uncertainty and physicist Jack Sarfatti's theory of retrocausality, that is the future determining the past, the experiment also indicated that if a test subject could place his or her future self in front of a couple of marshmallows, that would determine the test subject's actions in the present. A remote viewer might ask whether this test actually involved a subject's viewing the future with two marshmallows or one, while a proponent of Heisenberg or Einstein might ask whether the child in the future had already eaten two marshmallows and, thus, abstained from eating the single marshmallow on the table. The children's intent about the future governed their actions in the present, so as to make the present predetermined by the future. Or were the children simply driven by their imaginations: two are better than one.

There is no question that modern mediumship or modern paranormal has made progress not dreamed of a century or more ago. In a sense, the subject of the paranormal had evolved from séance to science. There are still plenty of skeptics, debunkers, and atheists. However, there are many more scientists today who are willing to explore the paranormal. Just as millions have reported near-death experiences, four out of every ten people claim contact with deceased loved ones. Science is not as narrow as it once was. Still, no one has invented anything close to Thomas Edison's idea for a spirit phone.

(December 6, 2016).

CHAPTER 15

Can We Talk to the Dead?
Edison vs. Tesla

Both Tesla and Marconi, the acknowledged pioneers in developing wireless transmission of electrons, believed that voices from the other side, voices from those who had died, could be picked up by radio signals because those signals were eternal. Tesla even believed that radio waves carrying signals from ETs from distant planets could be picked up by receivers at the right frequency, a precursor to the Search for Extraterrestrial Intelligence or SETI.

Perhaps one of the most astounding aspects of Edison's attempts to contact the dead was his belief that spirits of the departed wouldn't bother to move objects around or knock on tables to announce their presence. Reacting to the events he had observed during the Great Age of Spiritualism and the news of the Fox sisters' claims, Edison wrote his thoughts on life after death in a chapter of his diary called *The Realms Beyond*. (It was subsequently deleted from his diary, called *Sundry Observations*, by his editor and possibly by members of his family.) Edison, himself, did not delete *The Realms Beyond* chapter because he heartily believed in the science of quantum physics of his day. However, he did believe that such things as Ouija boards, tea leaves, spirit rappings, and other such forms of communication were a "waste of time." He wrote that any proof of the afterlife would have to be made on a "scientific basis," just as chemistry is proven, and not on the claims of individuals who might prey on the needs of others. For his part, Edison believed that science was dispassionate and not prone to the desires of those who had special agendas.[72] He had also learned from experiments he had performed throughout his life that scientific proof had to be repeatable and not simply a series of one-off events.

EDISON'S SCIENTIFIC PREDECESSORS: TESLA AND MARCONI

Edison focused a keen eye on his competitors, notably Marconi and Tesla, both of whom developed radio-wave transmission and reception devices. Both inventors might have picked up anomalous radio signals on their receivers early on in their experiments which, according to Tesla, were transmissions from the other side. Edison mocked Tesla's claims, but upon hearing them, set about to construct his own device that could ping the presence of the departed. In the *Lost Journals of Nikola Tesla*, Edison is quoted by his rival as having "privately believed that Tesla had managed to find the correct frequency to enable communication with spirits of the dead."[73] Edison was determined to discover Tesla's secret and be the first to get the spirit phone on the market. Tesla also believed

---

72 Edison, Thomas, *The Diary and Sundry Observations of Thomas Alva Edison*, ed. Dagobert D. Runes, (New York: Philosophical Library, 1948).

73 Swartz, Tim, ed. *The Lost Journals of Nikola Tesla*, (New York: Inner Light-Global Communications, 2000).

that Edison was able to communicate with the dead and wrote in his journals that "Thomas Edison had heard from other engineers that Tesla had been receiving mysterious voices and sounds over radio frequencies that were not conducive for the broadcasting of the human voice."[74]

Edison and his assistant, Dr. Miller Hutchinson, aggressively began to research the mechanics of a device to communicate with the dead, about which, Dr. Hutchinson later wrote in his own journal, "Edison and I are convinced that in the fields of psychic research will yet be discovered facts that will prove of greater significance to the thinking of the human race than all the inventions we have ever made in the field of electricity."

EDISON'S SCIENTIFIC BASIS FOR COMMUNICATION WITH THE DEAD

It was his goal as an inventor, Edison said, to create devices that would serve for others as a type of compass charting paths "upon scientific basis" into the unknown, in this case, the realm of the dead. But there was nothing supernatural about Edison's intent. Instead, he thought of the device as a "valve" which, just like a megaphone that increases the volume of sound, will increase the volume of the electronic registration of the phenomenon that it detects.

Edison said he was drawing a distinction between claiming that a specific human personality survived after death and an attempt to find some evidence of a spirit, a collection of entangled electrons representing what remained of the departed. He was actually redefining the concept of a ghost from something purely spiritual

---

74 Ibid.

into something electrical, the cohesive remains of a consciousness. The device Edison was developing might have been named a spirit phone, but it was more of a type of radar, a sensor device that would recognize the presence of a disembodied spirit, register it on a meter, and record its presence.

Relying upon his own extension of Einstein's special theory of relativity, Edison stated his opinion that life on earth was indestructible, which meant that life qua life, not the corporeal body but the essence of life, called "units of life," could neither be destroyed nor created. These units, Edison believed, were smaller than molecules and individual cells. In fact, he wrote that they comprised the ingredients of cells and were indestructible. Sir Roger Penrose has been reported as having described these as quantum units.

SPOOKY ACTION AT A DISTANCE

Here's where it gets interesting and comports with what will later become the theory of quantum entanglement: Einstein's "spooky action at a distance" theory. Edison said that life units can gather themselves together by the billions and in such a way as to make a human being. Simply stated, and almost in line with the thinking of his contemporary intellectuals that the unseen was reality and the visible was distortion, the life units that comprised a human being were submicroscopic. Like quanta of matter, Edison had to believe they were there and that the gatherings of them comprised something we could see, an individual human. This was an exciting concept because not only did it comport with the discoveries and theoretical arguments of physicists of the period, it reflected Edison's own interpretation of the biophysiology of that theory and the developing science of genetics. We also should be mindful of the fact that geneticists at that time were developing theories of eugenics, the deliberately focused breeding of human beings so as to create superior humans. We all know how that turned out.

Characterizing his "life units" as electrons, Edison was convinced that millions of life units comprised a human. Further, because they were electron-like particles, they were invisible. Not only that, but they carried electric charges, wherein each charge was a signal that hailed other signals, all of which possessed a form of memory. He theorized that because the human body can regenerate itself in a healing process, there must be some form of memory in the cells, a memory that probably resides in the electrons that comprise them. For example, he theorized that when a person burns his hand, the skin on the hand will eventually grow back in exactly the same way as it did before. How does the body know how to do that, Edison asked? The body can only do that if it possesses a memory of what it looked like before the injury. Thus, he postulated, this memory, if it resides in the electrons as life units, most likely exists independent of the body. And, if that's the case, then it should survive the body after death, especially because, according to Einstein's theory, matter cannot be destroyed.

A TAXONOMY OF LIFE UNITS AND THE ASYMMETRY OF THE HUMAN BRAIN

Perhaps predicting what would become known as the stem cell theory of manipulation of cellular development, Edison argued that there were "worker units," those that were the building blocks of what would become a human, and "master units," those that would govern how the worker units formed up. Thus, ordering the workers into functional structures was the work of master units, which, he also presupposed, might reside in a specific location of the brain, an area, he said, that is the "seat of personality." He said this master personality area was likely the Fold of Broca, which neurologists now know is where the hard-wired structure of language is located. This area, in a right-handed person, is situated on the left hemisphere of the brain and is also the seat of predicative logic and reasoning.

The human brain is asymmetrical, that is to say, the two hemispheres have generally different functions. The left, in a right-handed person, is the seat of logic and ordered behavior and generally controls or supervises the more creative and non-ordered ideations on the right hemisphere. This is an oversimplified description, which overlooks the brain's complexity and plasticity. The two hemispheres are joined together by a bundle of ganglia or neurons known as the corpus callosum, a kind of biological heavy-duty telephone cable or T1 pipe that connects the brain's motor and cognitive functions.

What we know about human development tells us that, as prehistoric human beings, we began to utilize fire to allow clans to own the night, protecting themselves from wild animals, warming themselves, and cooking meat. Civilization of affiliated clans began to evolve into hierarchal pack structures. Humans began to use tools, and the tools they used, worn more on one side than the other, indicated what we call "handedness," the favoring of one side of the body over another. This indicated that the two hemispheres of the brain were becoming specialized; structured language was part of that specialization. Today, this is also known as predicate organization. We remember predicates from elementary school grammar, as in the subject and predicate of a sentence. The subject holds the predicate much like a hand holds and wields the tool, much like the logical structure of an argument holds the premises and conclusion together. In this way, Edison was way ahead of his time in that his theories of where the master life units resided would later prove to be the explanation of predicative reasoning.

COHESION OF LIFE UNITS AND THE ETERNITY OF LIFE

Do these master units stay together as a collection of entities after the death of the body or do they dissipate? That was the question that perplexed Edison in the early 1920s. If the collective units that form a cognitive human being break up, then there would likely be no life after death. The human consciousness just ceases to exist as an organized group of units. In this way, eternal life, which is the belief system of major religions, would not mean the eternal life of an individual, but an impersonal constellation of free-floating electron particles without a sense of personality.

Edison said he hoped that human personality did persist in a cohesive electronic state after death. This was the primary reason he was creating his apparatus, to measure the electronic pulse, the registration of an electronic presence, no matter how slight, that would assure him of the existence of an eternal life.

The life units that Edison hoped to prove lived after the death of the body also helped define what some religions called reincarnation. This was another radical thought that Edison was working through. He said that the same set of electronics that might explain the persistence of life units after death might also explain how these life units reformed themselves into memories in new bodies. That would explain how a person might have memories, or "flashbacks," of another life. It would be a completely scientific explanation of reincarnation.

Edison theorized that we do not actively remember things. It's the life units, the "little people" in our brains, he said, that provide us with memories. And these little people reside in the Fold of Broca. In this he was oversimplifying, because we now know that human memory is a complex process that resides across the brain from short-term or immediate memories to long-term storage in a hierarchical organization. Deep or very long-term memory may reside in part in the animal portion of our brains, in which the fight or flight emotion is located. Much of the modern biology of memory was theorized by the psychologist Donald O. Hebb, who is considered by many to be the father of neural network theory, the basis of today's computer memory technology.[75] Professor Hebb, who hypothesized that "neurons that fire together, wire together," also worked for the CIA during what became known as MK-ULTRA, the manipulation and alteration of an individual's memory.

MEMORY AND FALSE MEMORY

Nineteenth-century physician and anthropologist Pierre Paul Broca showed that much of what we call memory—he was probably referring to short-term memory—resided in a strip along the left frontal cortex. These "little men" kept memory records for us, Edison wrote. Just as in motion pictures, the little men simply watched, recorded, and replayed. Notice how much Edison's conceptions of what Broca theorized were governed by his own inventions. Edison pushed this analogy further by saying that what we see, our vision, is really a series of snapshots or stills of reality that hit the optic nerve and are reassembled in the brain as motion pictures. As we get older, we may tend to forget those

---

75 Hebb, Donald O., *The Organization of Behavior*, (New York: Wiley, 1949).

things that have not made great impressions on us because our memory records get used up. When we're young, however, the slightest impressions can fill our memory cavities because they are empty at birth. These childhood memories, being the first ones in, tend to linger throughout our lives in such a way that even old people may still react to them as if they were current. Insofar as the brain is concerned, they are.

This phenomenon, Edison explained, illustrated a psychological phenomenon that can play out in dreams. Imagine dreaming of your former schoolroom and the proportions of your own body to your desk, your chair, or the windows of the classroom. In your dreams, you are replaying the memories you stored as a child. Now were you to return to that schoolroom as an adult, everything would suddenly be wrong. The desks would be too small, you wouldn't fit into the chair, and you would be the wrong height for the window. All of this might make you uncomfortable. You might ask yourself why, but the answer is that your current experience in that classroom doesn't match up with your memories. As Edison explained it, though the memories may remain with you as they were impressed upon you as a child, your body changed as you grew up. The little people in your brain do not change. Your body does change, and thus there is disconnect between the little people and the person you are.

This phenomenon of incongruent or dyscongruent memory, Edison explained, also illustrates how memories update one another. Again, using analogies from his experience as the inventor of the recording and communications industries, Edison argued that the little people who store memories have to compare what memories reside in the brain with new impressions that come in. Thus, there is a constant relay interchange that takes place, a set of signals, between stored memory and new impressions. We use clues to trigger stored memories that we have buried. When we remember something from the past that's triggered by something from the present, it's as if the memory comes crashing into our consciousness. That's because the little people who hold the memory shout that memory back into the person's consciousness. This is also one of the modern theories behind post-traumatic stress disorder or PTSD, in which an event in the present can trigger the emotions of the past and govern a response. Where Edison's theories about the little people who store memories and his life units coincide is in his theory that the little people, the life units, do not disappear after the person dies. Rather, they linger.

Thus lingering, the little people can inhabit other bodies so as to preserve memories from a former life. In fact, Edison writes, these memories are carried forward through a succession of lives so that the memories of a person are themselves immortal. Not only does this explain, for Edison, reincarnation, but it also explains his theory of immortality. This was yet another reason behind Edison's drive to prove his theories correct with his invention of the spirit phone.

THE TRANSMISSION OF LIFE UNITS

It is here that Edison faced a quandary. If he were correct and the recollections of the little people remain throughout a succession of lives, then each generation must gain from its own ancestors so as not to make the same mistakes. Given this reason, Edison surmised that the Franco-Prussian war of the late nineteenth century should not have been repeated in the Great War, World War I. But Europe did not learn. How could this be if Edison's theory about the permanent existence of recollections were correct?

For an answer, Edison drew on the theory of collective consciousness by psychiatrist Carl Jung, and argued that some, but not all, of the recollections carry over to succeeding generations. This, he said, accounted for "inherited wisdom." For example, how does a baby know to close its hands around an extended finger from its parent? Although today we might argue that that is a reflex action based on instinct, Edison argued that it is part of a shared inherited wisdom, something that remains—call it genetically—from one generation to the next. In this argument, Edison was also way ahead of his time, even though we've replaced his terminology with our own.

What he called "inborn traits" were actually "recollections of earlier experiences that the little people brought along with them." And he used examples to illustrate that different cultures pass along their cultural norms to succeeding generations without actively teaching those norms. Instead, each generation's observations and inborn traits are incorporated into those of the next generation. These recollections are what create consciousness, invigorating the human body with both sentience and sapience.

From a materialistic perspective, Edison perceived the individual human as a "machine," in his words, that was composed of millions of life units or little people who comprised the sum total of the person's memories and emotional responses based upon those stored memories. Just like an extension of both a Freudian theory and a Jungian theory, the individual person's reactions and motivations were governed by his or her memories, which were actually life units, electrons. But like Jungian theory, these life units were shared among individuals, hence the idea of a collective human belief system. Moreover, from an Einsteinian perspective, because the life units were matter and could not be destroyed, if they remained together, joined by a spooky attractive force, then it might be possible to perceive them as a personality surviving the death of the body machine. In other words, if Edison's spirit phone worked as he intended it to, then it would prove that life does not cease to exist at the point of body death. The apparatus, Edison theorized, would also help him prove his theory about the mystery and origin of life, not just about death. Edison argued that when we see an individual human being, or an animal, or a plant, we see that individual as a unit. This, he believed, was false. Because life units can exist independently of the human body, they exit that body upon the body's death and pass into another habitat, almost like a virus that spreads to a new host. From the perspective of reincarnation, Edison wrote, they might become the animating force of another machine.

MIGRATION OF LIFE UNITS

The principle of life units also gave rise to another Edison theory concerning what he formerly lamented as the "cruelty of nature." But nature was not really cruel, in his opinion. It was merely the action of some life units transferring to another habitat upon the destruction of the machine they originally inhabited. He used the example of the shark that eats the cod. The cod may die, but it is only the flesh. The life units that make up the cod simply inhabit, cojoin, the life units of the shark. Hence there is no death of the indestructible life units, only a recombination. This theory also explained much Native American lore that one is what one eats. If one eats the heart of the wolf, one becomes like the wolf by incorporating its spirit. If we look at the spirit as Edison's life units, there is a materialistic basis for what would be Native American spiritual beliefs.

In his theory of life units, Edison mused upon the smallness of the units he was describing and came to the conclusion that if we view them as electronic particles, whose existence had already been proven, then there is no longer a question about seeing these life units with a microscope. We know they exist because electricity exists. This was another aspect of the rationale for his spirit apparatus.

Edison was so convinced of his theory of life units that he believed he could also demonstrate proof of the subconscious mind, which was promulgated by Freud at the turn of the century. Edison suggested that notions of good and bad, therefore, were really determined by the nature and number of the master builder life units, those that control the life units. But where do they come from? Do they come from inside the sperm and the egg in which a nucleus is fertilized, thus giving rise to the offspring? For Edison, and this also explains the rationale for his spirit phone, the answer was a flat-out no.

Because life units can live outside the body and because, in his words, they can "prowl about the atmosphere," he believed they existed independently in a form of ether. Does that mean that the origin of life, because the electrons can neither be created nor destroyed, is outside the body? Perhaps life came to earth from outer space. He wrote that, as the earth cooled from its fiery beginnings in the solar system, the units of life came to earth from another, more advanced planet. This was a theory advanced by Lord Kelvin, among others, and was probably part of the scientific scholarship that influenced Edison. It is probably this reasoning, something completely non-Christian, that motivated Edison's descendants and successors in interest to delete this chapter from later editions of his published diary. Edison's idea of migrating life units, however, was far ahead of its time and preceded Crick's and Watson's theory of panspermia. If we someday in the very near future find fossilized quasi-terrestrial life forms on Mars, for example, or even among the hydrothermal vents on the Saturnian moon Enceledus, perhaps not only Crick and Watson, but Edison, too, will be proven correct in this theory of the migration of life units.

EDISON AND WILHELM REICH

These life units were eminently adaptable and upon reaching our now-cooling planet, simply adapted to the chemistry of our world and our atmosphere, thus creating life on planet Earth. This postulate involves an understanding of how these life units travel. If they are bound only to travel through an atmosphere, then they are limited by the confines of that atmosphere. However, if they can travel through ether, the popular term also used by Wilhelm Reich to explain his theory of biological energy or Orgone, then they are not confined by our atmosphere.

Reich was a psychiatrist who had been a member of Freud's circle of confidants, colleagues, and students, and sought to explain the nature of how neuroses are transmitted, almost like pathogens. He believed in the existence of a conducting device in what he called the ether, and named that conducting medium *Orgone*. From Edison's ideas concerning the transmission and travel of life units through ether, it seems plausible that he was influenced by Freud's and Reich's ruminations on how the unseen forces that govern human behavior and comprise character and personality migrate.

For the sake of argument, and assuming that Edison's theory of life units had plausibility, we ask, and he asked as well, are life units not only indestructible but immutable? If immutable, how does that account for the diversity of life on earth? Edison suggested that because life units are adaptable to their environment, as the environment changes, or evolves, so do the life units that comprise the separate organisms.

216

This adaptability enabled him to explain that although an elephant is an elephant is an elephant, the wooly mammoth is not the same elephant found in Africa or Asia. As the earth warmed from the Ice Age, a wooly outer layer on the elephant became unnecessary. Thus, the life units adapted to the new climate and the elephant lost its wool. How Darwinian is that? We can see via these arguments how Edison was a thinker of his age who amalgamated the theories of the day to advance theories that are only now being taken as good science. Having established his postulates about the nature of life units, their migration through space to form life on earth, and their indestructibility and ability to exist in cohesive swarms absent the corporeal body, Edison turned to his next important principle—that we can communicate with the dead.

CHAPTER 16

The Spirit Phone Apparatus: The Principle and Design

"I cannot conceive of such a thing as a spirit," Edison once wrote in his diary.[76]

Edison didn't think invisible spirits could make raps on walls or tilt tables. However, although he did not think that what we call "personalities" passed to "another sphere" upon death, he did think they could be contacted by an apparatus. This apparatus had to be delicate and sensitive to the slightest perturbations in the electronic envelope that surrounds us all, his version of the Force or the Vril. Its purpose would be to register the presence of these perturbations to see whether, against the background noise, something cohesive stood out on a meter. Moreover, Edison believed that if he could construct such a device, it might not only be able to identify the existence of a cohesion of electrons, but allow that cohesion, assuming that that cohesion actually formed

---

76 Edison, Thomas, *The Diary and Sundry Observations of Thomas Alva Edison,* ed. Dagobert D. Runes, (New York: Philosophical Library, 1948).

a personality, to "express itself" to those tracking it. His apparatus would be based upon science and not subjective impressions of clairvoyants or mediums.

BRINGING SCIENCE TO MEDIUMSHIP

Edison confessed to future readers that it was not the idea of life after death in principle that he was challenging, it was the unscientific and crude methods of communication employed by self-proclaimed spirit guides that he rejected. Poltergeists, for example, seeking to make their presence known to the living, would simply be wasting their time knocking things about. Edison's method would provide a clear and objective means of communication, like a telegraph or telephone, open to no misinterpretation. The field of psychic interpretation would grow in legitimacy by leaps and bounds once the science of electronic communication were applied to it, he wrote. And in this, again, Edison was blending the cultural trends of spiritualism and scientific industrialism into a single piece of electronics whose proof of concept would, he believed, satisfy practitioners in both fields.

The apparatus, Edison believed, would have the effect of magnifying human effort to open a doorway—Edison called it a "valve"—into another realm. Like a modern powerhouse, a generating plant that moves electricity along a grid, his device would act as a turbine magnifying the input force, in this case the appearance of free-floating electrically charged life units, to a meter that would display its presence. Once registered on a meter, the event would be recorded. Hence, Edison asserted, the user would have a record of the appearance of the spirit via an apparatus that would return scientifically repeatable results every time it was used. Just imagine, using science to prove about the hereafter what heretofore was completely unscientific. By merging the dominant cultural, scientific, and intellectual trends of the times, Edison was more than a man of his time, he was displaying his intent to become a man of the future.

EDISON'S DESIGN VERSUS TESLA AND MARCONI

Whereas Tesla and Marconi believed they could construct receivers that would pick up radio waves from the other side, Edison thought that the presence of life units would provide impressions from the other side. This was a fundamental difference between the theories of electric signaling. Tesla's theories of voices from the other side were predictive of the modern practices of listening for EVP, or voices of the dead, while Edison's theory was predictive of "spooky action at a distance" or quantum entanglement.

STEPS TO CONSTRUCTION OF THE SPIRIT PHONE

Edison's next step would be to construct the apparatus. Electricity would generate the power to project a tightly constrained photon beam at a photoelectric cell that would register the impact of those photons on a surface medium. The photoelectric cell was connected to a meter that would then register whatever passed through that beam, so as to distinguish the effect from any background radio waves, and record the interruption of the beam. So far, so good, but Edison would still have to know that a cohesion of life units was actually present and could be summoned across the beam instead of sitting around and waiting for a random spirit to present itself for inspection and registration. In this, he claimed, he was fortunate in that a person with whom he had been collaborating on this device had unfortunately just died. But inasmuch they had been working together, Edison felt that that spirit of the just-departed should be the first entity to register himself upon the new apparatus.

Cautious about the capabilities of the apparatus, Edison characterized his work as giving psychic investigators a means of scientific proof of the afterlife, just as a microscope gave biologists proof of the existence of organisms they could not see with the naked eye but knew were present. Microscopes used optics to make what was unseen seen. In this way, Edison's apparatus also used optics, light, to make what is only perceived psychically, registered electronically.

EDISON AND GENETICS

Edison also articulated another theory in regard to the structure of electronic life units. And we also have to be aware that he is making this argument at a time when the entire study of genetics and an organized human genome was being researched. In fact, the 1920s in Europe and America was a time in which not only was genetics being studied, but some biologists were arguing for the practice of eugenics to improve the human race and epigenetics to explain how certain seemingly acquired traits were passed from generation to generation.[77] Edison, although not arguing for the theory of epigenetics, did come up with the theory that the organization of human beings into hierarchical structures mirrored the way life units organized themselves into hierarchical structures in a human being. In other words, in arguing that the organized structure of human society, with its managers and workers, is a reverse projection of the organization of the life units within the individual

---

77 cf. Pottinger, Francis, *Pottinger's Cats: A Study in Nutrition* (Lemon Grove, CA: Price Pottinger Nutrition Foundation, 1983).

human, Edison was predicting the theory of fractal mathematical representation, a form of scalability or a recursion theory in which ontogeny—the organization of the one—follows phylogeny—the organization of the many.

Finally, in arguing for the theory of what spiritualists call migration of the soul, Edison stated that life units are indefatigable entities, always seeking the opportunity to work. Plus, they are indestructible. Thus, when the body machine dies, the life units seek out another body machine to inhabit. In other words, there is a migration of the soul, not on a spiritual level, but on a bioelectrical level for which hard science can account. Moreover, because there is a fixed number of life units, they recirculate, inhabiting human flesh and blood over and over again, only in different bodies. Hence, there is an electrical theory to support Jung's contention of a collective unconscious of shared symbology and ideas that resonate through cultures.

Finally, Edison turned his attention to the nature of personality itself. After all, if life units recirculate from biological machine to biological machine, how do different personalities form up out of the same soup of life units? His theory presupposed that the life entities recombine in different proportions when they migrate. They, like Einstein's mass, never go out of existence, but like swarms of bees from different hives, they can form new hives and new hierarchical structures, and assemble different combinations to form different personalities. Thus, Edison suggested that by registering the presence of these cohesions of life entities, his apparatus would prove not only the existence of life forces after death, but, by allowing the electrons to communicate with the operator of the apparatus at the other end, might be able to display lingering personality traits in a type of Morse code. Seems like a stretch, but to

Edison's mind it was a worthwhile experiment, given the theories of his day.

Now it was time for building the apparatus, something that Edison described in his journal as the "most sensitive apparatus I have ever undertaken to build, and I await the results with the keenest interest."[78]

---

78   Edison, Thomas, *The Diary and Sundry Observations of Thomas Alva Edison*, ed. Dagobert D. Runes, (New York: Philosophical Library, 1948).

CHAPTER 17

The Device and the Skeptics

The first official independent article about his spirit phone, one not written by Edison, appeared in the October 1933 edition of *Modern Mechanix*. In that article, Edison described to an interviewer his first public experiment with the apparatus in 1920, and also explained how the device worked. In that experiment, Edison invited only those he trusted, what *Modern Mechanix* referred to as his trusted "mystic clan" of scientists. Although we do not know exactly what happened that night, except that the magazine reported that the experiment did not reveal the presence of any spirits, the story goes that Edison had constructed an arrangement of a photoelectric cells that was supposed to register a narrow beam of light projected upon it from a lamp set up a short distance away. The room was dark, so that only the beam of light could be seen by the onlookers.

When the beam of light struck the photoelectric cell, the cell transformed the beam into an electric current wired to a meter. Any interference with the beam would register on the cell, and a needle would move on the meter in response to the interruption of the beam. Although the apparatus seemed basic, it was ingenious in theory because it presupposed that something unseen would still be perceived by the stream of photons hitting the cell. For the experiment to work, though, Edison had to make sure that he had a cohesion of life units, a spirit, from a departed person crossing the beam. For that, he had to rely upon the claims of spiritualists, mediums, who promised him that they could summon the departed. Though Edison was a staunch denier of the abilities of mediums, how else would he be able to find subjects to cross the beam of light unless he had a way to summon them in the first place? Thus, the premise of the experiment, if not of the device itself, was to amalgamate the materialistic—a cohesion of entangled life units—with the spiritualistic—an actual séance to summon the departed.

The experiment began. The mediums commenced whatever trancelike induction they needed to reach out to the spirit world to see who was navigating nearby in the ether. This was less like magic and more like a true psychological event. As they performed the machinations they needed to reach the spirit world, the scientists watched the meter to look for any interruption of the stream of photons.

Hours passed as the spiritualists kept reaching out with their invocations. The dial on the meter did not budge. As the night wore on, it was becoming clear the spiritualists were not able to summon any entities. The scientific skeptics among them thought there were simply no entities to summon in the first place. As ingenious as Edison's concept might have been, and even if there were spirits of the departed nearby, they simply could not be coerced into taking part in the experiment. For whatever reason, the experiment was not working. Thus, there were at least four basic theories concerning his failed experiment:

1. The mediums could not summon any spirits.
2. There are no such things as life units of the departed.
3. Even if there were spirits and life units present, they simply did not register on the photoelectric cell even if they interrupted the beam.
4. Even if there were spirits present and even if they could be summoned, the premise of the spirit phone was faulty and there was no physical matter to interrupt the beam.

For Edison, the failure of his experiment was not a devastating shock. First of all, he was used to failure. He often said that failure simply meant that he could eliminate one process or type of engineering project from his theory and move on to the next. Therefore, he didn't abandon his theory that life units remained after the death of the flesh. It simply meant that whatever his device was supposed to detect, either the spiritualists weren't capable of doing what they said they could do, or the machine itself needed a different method of detecting what he believed present and active in all of us.

Edison had argued that he had proof of the existence of life units that could reform even after the flesh was destroyed. He used the example of a fingerprint. He wrote and told others that the fingerprints of an individual were unique. No two people had the same fingerprint. And if the tip of the finger upon which the print existed were burned so that the fingerprint was obliterated, when the scar tissue finally healed and the fingertip returned to its normal state, the same fingerprint would reappear exactly as it had been. To demonstrate this as a proof of concept, not only of the healing of the flesh but of the presence of indestructible life units, Edison first took his own fingerprint to get a control specimen. Then he burned that fingertip severely. When the scar began to decay as the burn healed, Edison pulled off the tissue and had another fingerprint taken. Lo and behold, he was able to show that the new fingerprint was exactly the same as the original one. The life units repaired the skin in exactly the same way. Hence, for Edison, this was proof of the concept that life units existed and were indestructible. Therefore, it was not the theory that was at fault, it was either the design of the device or the inability of the mediums to summon the spirits.

EDISON AND EXTRATERRESTRIALS

Edison gave his interviewer at *Modern Mechanix* another example of his life unit theory. Suppose, he asked, that an extraterrestrial being from another world had eyes that were incapable of perceiving anything smaller than large structures. It could see buildings and large statues, but nothing smaller. Now suppose that same extraterrestrial came to Earth and could see the Brooklyn Bridge, but not the people walking across it. Then what if, somehow, something destroyed the bridge and the extraterrestrial could see it was no longer there. Now suppose New York construction crews rebuilt the bridge. To the extraterrestrial, the bridge simply reappeared, just like Edison's fingerprint after he had burned it off. Might the extraterrestrial then believe that something unseen had rebuilt the bridge, even though it did not see that process take place. This, Edison said, was exactly what happened with the life units. They work invisibly, but all we can see is the results of their work. By perceiving their results, he argued, we therefore have a proof of concept that life units exist and shape, form, and animate the human machine and all other forms of life on Earth. Clearly, Edison, ahead of the wave of twentieth-century biological science, had come up with a theory that would later be proven to be true: gene theory and cellular memory.

Unfortunately, Edison's life-unit-theory experiment could not account for the presence of life units after death of the entire body, nor could it account for the possibility that an electronic device could perceive the presence of these swarms of life units. This was perhaps the reason why Edison never fully revealed to the public the work he was doing on his spirit machine, only sharing it with a few trusted colleagues and those he believed honest enough to bear witness to its success, were that success to have been proven. But that did not stop skeptics from dubbing his experiments as a ghost phone or telephone to the dead.

PURITAN AMERICA

Despite the failure of his experiment, Edison still clung to the belief that life existed after death and that science could prove it. But America was still a Puritan country in the 1920s. Ghosts, spirits, and the idea of communicating with the departed were all pushed to the margins as industrialized America invested in itself with money it didn't have. By the late 1920s, although Edison's energy and capacity for creativity might have been in decline, he was still a legendary industrialist as well as an inventor. As public figures, those who sought profit from the industries he created wanted no part of spiritualism, and told him so when news of his spirit phone experiments leaked out. Therefore, all records of his experiments with the spirit machine were wiped away.

There is another issue that might have arisen regarding Edison's theories about life units, one that involved how these units arrived on Earth. Edison wrote in his diary (and these remarks were stricken in later editions) that he believed it possible that life units were also carried to earth by some extraterrestrial force, a precursor to the theory of panspermia. Were that the case, it would have given a purely materialistic, albeit extraterrestrial, theory for the creation of life on Earth and possibly elsewhere in the universe. For the head of a major company such as General Electric and a child of devout Methodist parents, this was pure heresy. Not only was Edison writing something that flew in the face of the Judeo-Christian story of creation; he had traversed into the realm of science fiction. For the most practical of inventors, this was a great leap into what his family considered to be fantasy. All of this was before Edison claimed that he had what we now call a near-death experience.

EDISON'S NEAR-DEATH EXPERIENCE

On his death bed in 1931, Edison lapsed into what his family thought was a coma. His breathing slowed and those at his bedside thought that this was the end. But suddenly, Edison roused himself, opened his eyes and, with the lucidity of a healthy man, spoke. He told his family that he had been to the other side. His voice was full of hope. He had seen what it was like after the death of the body, what the hereafter was like: a blissful place. But the true happiness in Edison's voice in his final moments was that he knew for certain, knew in a way that no scientific experiment could possibly prove, that there was consciousness after death. He had discovered, at the very point of death, we don't die.

CHAPTER 18

What Happened to Edison's Last Invention?

In 1920, Thomas Edison revealed that he was building a machine that he believed could contact the dead. He also revealed that he had demonstrated his design to scientists and employed the expertise of mediums to summon the departed to test the machine. In Tesla's journals, he referred to Edison's building a machine to contact the dead. With all evidence, albeit anecdotal, one question remains, what happened to the spirit phone?

TIMELINE

1920: Edison demonstrates the spirit phone to scientists and mediums

1931: Edison has near-death experience

2013: Sir Roger Penrose explains his theory of the quantum theory of consciousness

WAS IT A JOKE OR HOAX TO CONFOUND TESLA?

There are some Edison scholars who argue that the spirit phone never existed, that it was a hoax perpetrated by Edison upon his gullible interviewers at *Modern Mechanic* magazine. Even though the apparatus seemed simple in design (a motion-picture projector-like device that focused a narrow beam of light upon a photoelectric cell wired to a meter to register any interruption), the fact that the device was never officially put on display or demonstrated to the public made it seem as if Edison had either failed or was making a joke for his fans. But Edison did not joke about his inventions. In fact, a dour Thomas Edison did not joke at all, except to laugh at Tesla's demand for payment of $50,000 for his repair of a generator. Whatever his humor might have been, when it came to inventions that he believed would satisfy a consumer market, Edison was strictly serious. Accordingly, we have to wonder not only whatever became of the device, but why was Edison's chapter on the spirit world and life after death excised from his diary when it was published by his family?

As Edison's health declined in the late 1920s, his businesses were prospering. The recording industries were making technological advances, such as converting cylinders to disks that could be pressed by the thousands; the motion-picture industry now had talking movies, and studios were being rewired for sound by pioneers like NBC founder David Sarnoff in his work with Joe Kennedy, Sr., at RKO; municipal power grids were being expanded to more and more cities; and the transportation system was being electrified. Edison's General Electric Company, GE, which had begun in 1890, by the start of the First World War in 1914 had become a defense contractor, manufacturing warplanes for the Army Air Force. By the late 1920s, Edison had even brought early experimentation in plastics to his company. In short, Edison had gone from being a lone inventor to the head of an invention factory, to a corporate personality with far more at stake than just a single invention. Edison had joined the ranks of entrepreneurs such as

Judge Thomas Mellon, Andrew Carnegie, George Westinghouse, and John D. Rockefeller.

INVENTIONS AS EDISON'S PRIVATE PASSION

Even though Edison had brought the age of electricity to consumer America, he remained an inventor at heart, pushing the envelope to find new pockets of need for inventions. However, by the 1920s, the private Edison the inventor and the public Edison the corporate businessman found that sometimes the latter compromised the former. By the time he had become corporate, it seemed that Edison was not allowed to fail. There were simply too many people, investors and business associates, on the line with him. When one is the personality upon whom the value of publicly held shares of stock depends—consider the personas of Bill Gates and Steve Jobs—that person must walk a fine line between safe continuity and the risk-tinged potential of reward. Look at the fights between Steve Jobs and John Scully: the former, who founded Apple, arguing that the company's future lay in the development of the Macintosh, while the latter saw the safe income stream of the Apple II as the company's life blood. That same type of conflict might have torn at Edison internally. He wanted to determine whether the dead could make their presence known to the living, but he also had to maintain his role as the corporate celebrity upon whom entire industries depended. Therefore, it would not be surprising that his failure to develop an apparatus to communicate with the dead would be seen as a detriment by those who depended upon his renown, and thus all records of that failure had to be expunged. And we're also assuming, for the sake of argument and with lack of any evidence to the contrary, that it was a failure.

But what if it wasn't?

WHAT IF THE SPIRIT PHONE ACTUALLY WORKED?

What if, upon a subsequent night with a different group of mediums or trance channelers, Edison managed to get a registration on the dial of his apparatus? We know that he described exactly how he would build it and we know the principle behind it. Therefore, suppose that there were spirits, or cohesive life units of the departed who had crossed the beam. Would Edison have brought those results to the public or would the board of directors of General Electric or of any other Edison-related businesses have advised him against doing so? The thought that a person could scientifically prove the existence of spirits, as opposed to a faith-based belief, would be like a Dr. Frankenstein creating life from the dead in a laboratory. Just imagine what would have happened had the founder of General Electric suddenly announced that he could communicate with the dead. Would it be a boon to his company or would organized religion come down upon him like the pillars of Sampson? This would have been choice faced by those responsible to the GE stockholders.

Although we have no proof one way or the other, are there any similar situations we can refer to as examples of how advanced research into areas considered paranormal were suppressed by the government? The answer is yes, there is an example that took place over a decade after Edison's death in 1931. It was in 1943, after the death of Edison's old rival, Nikola Tesla.

TESLA'S ANTI-GRAVITY NOTES

To understand the import of what took place in the final years of Nikola Tesla's life, and his attempts to get the US military to fund his research into advanced weaponry in the years after the beginning of World War II, we have to look at an attempt that Tesla made to offer advanced weapons to the Navy at the advent of World War I. At the time, the person advising the Naval Consulting Board was Thomas Edison. We have already covered what writers such Margaret Cheney wrote in regard to the "battle of the currents," when Edison and Tesla argued over the efficacy of direct current versus alternating current.[79] And we know that even before Tesla came to America, he had proclaimed his wish to work for and with Thomas Edison, the master of his age. However, as Margaret Cheney documented, the story of Tesla's proffering advanced weapons technology to the Naval Consulting Board is not well known.

As Europe was gearing up for war, Tesla proposed a number of advanced weapons to the Naval Research Board. Perhaps the most important were directed-energy weapons, beams of electrons fired with such high intensity that they could be targeted quickly, didn't have to be reloaded, and could fire with pinpoint accuracy. These are the very types of weapons proposed today for the Sixth Generation fighter now on the drawing boards at Northrup Grumman and for the next generation of warships. Used at sea, directed-energy weapons would enable a single warship to engage multiple targets simultaneously from any angle and position. Today, just to prove the viability of Tesla's theoretical use of particle-beam weaponry, the Army has developed directed microwave weapons for the purposes of crowd control which, when aimed at a human target, heats up the person's skin to the point of intolerable pain, thereby rendering a hostile person immobile or sending him or her into flight.

The other superweapons that Tesla proposed were robotically guided torpedoes and warships. And we have also seen the flaw in

---

79 Cheney, Margaret, *Tesla, A Man Out of Time* (New York: Touchstone, 2001).

early radio-controlled devices that could not cloak their frequencies so as to prevent them from being jammed by the enemy. It took Hedy Lamarr and George Antheil to fix that problem, but only decades later. The robotically guided artificially intelligent underwater drone is actively undergoing development today by the US and Russian navies. Tesla had demonstrated his radio-guided robotic devices at Madison Square Garden in 1898. Even today he is known as the Father of Robotics. Tesla's radio-guided warship and torpedo might have given the allies an early advantage in the North Atlantic war against German U-boats. But it was not to be.

Tesla's weapons concepts, although more in the planning than physical development stages, nevertheless represented the brilliance of his theories of physics and electrical engineering.

Always looking for investors in his technology, Tesla had bounced them off financiers J. P. Morgan and George Westinghouse during his most productive years, but he usually managed to snatch defeat from the jaws of victory because, he claimed, the people he was selling to never really understood his ideas. For example, unlike his nemesis Thomas Edison, who understood market needs and developed inventions to fill those needs, Tesla thought too far ahead. He foresaw needs that would come to be in the distant future and developed technologies for them. But, as George Westinghouse once told Tesla, who was extolling the promise of free energy, why would anyone invest in something that would make no money for its investors in a market no one understood?

WIRELESS ELECTRICITY

Perhaps the best example of Tesla's optimism when it came to his inventions was the wireless transmission of electricity. Who needed wires, he believed, when he would harness lightning in a bottle, direct it toward receivers, and light the lights of cities in the same way he and Marconi had developed wireless communication? But, as Westinghouse explained when Tesla asked him to invest in his wireless transmission technology, what was the point of transmitting electricity wirelessly without meters? Why give it away for free? Free energy was a pipe dream, because how could an industrialist make money from it? This was the core of Tesla's problem when it came to raising money. Yet today, according to Cathal O'Connell's February 24 article in Cosmos, researchers at the Wireless Systems group at the Walt Disney Company's Disney's Research division, developed a method of using magnetic fields to transmit power to electronic devices wirelessly. It was Tesla's research in power transmission, Disney scientists said, that inspired them to use magnetic fields to transmit energy.

His financial problems were also at the forefront when he approached the Naval Board with his advanced weaponry. Edison rejected Tesla's plans for the wireless-directed energy-beam weapon and robot-guided torpedo as not feasible. It was his revenge, some historians say, for Tesla's winning the contract to light the turn-of-the-century Chicago World's Fair with alternating current instead of Edison's direct current.

TESLA'S LAST INVENTION: ANTIGRAVITY

By 1940, Tesla was living alone in a room in Manhattan's New Yorker Hotel. Europe, again, had gone to war and Tesla, again, had an advanced technological weapon, an antigravity device, that he was trying to sell to the United States Armed Services. Nikola Tesla theorized that electrically excited quartz crystals, at the correct frequency, could be directed to levitate an object to the point where it would overcome the force of gravity. When applied to weapons technology, Tesla said, antigravity would be the device to end all war, freeing humanity from the confines of the earth. In fact Ralph Ring, who appeared on History Channel's documentary series "UFO Hunters," said on camera that he was once solicited by businessman and engineer Otis Carr, who claimed he had worked with Tesla. Carr asked Ring to be a crew member of a flying craft, an antigravity device that Carr had built according to Tesla's antigravity notes. Ring explained that without any sensation of motion whatsoever, the craft that he was on transported itself ten miles. Ring could not explain the technology behind it except to say that Otis Carr had been friends with Tesla and understood his theories.

At the outset of World War II, Tesla was so enthusiastic about his antigravity device that he tried to obtain a grant from the United States Department of War for weapons development. But he was turned down, this time not by Edison, who had died over a decade earlier, but by the Department of War itself. Disconsolate, Tesla turned his attention to the Soviet Union, at that point facing the imposing might of Hitler's *Wehrmacht*. Anxious for new weaponry, the Soviet Union saw the practicality of an antigravity device and awarded Tesla a grant of $25,000 for the development of the apparatus. Unfortunately, Tesla was too ill to get the job done, and passed away in January 1943.

Tesla's grant from the Soviet Union did not go unnoticed by J. Edgar Hoover, the director of the Federal Bureau of Investigation. After Tesla's death, FBI agents raided his hotel room and secured all of his notes, including his antigravity plans and schematics, which they turned over to the custody of the Office of Alien Properties. After the war ended in 1945, the newly formed country of Yugoslavia demanded the return of Tesla's notes for a museum they were building in his honor in Belgrade. The Office of Alien Properties turned over the bulk of Tesla's materials to the new museum, but held back, perhaps as a matter of national security, Tesla's notes on antigravity. Instead, these notes were sent to General Nathan Twining at the Air Materiél Command at Wright Field outside of Dayton, Ohio. Twining, as mentioned earlier, was the general apprised of the UFO crash at Roswell, New Mexico, in July 1947. Debris from that alien crash was sent for analysis to Wright Field, where in spring of 1948 a class from the National Air War College not only handled the strange material retrieved from Roswell, but saw and communicated with the live alien who survived the crash.[80]

GOVERNMENT SECRECY A POTENT FORCE

The point of this Tesla anecdote is not to reinforce the story of the crash at Roswell. Truth can stand on its own over time and often needs no reinforcement. Rather, it is to point out that when the government seeks to keep an invention from the public, it can do so without too much of a cover-up. After all, even though we know that Tesla believed he had developed an antigravity device, and we know he had developed radio-guided robotic devices and directed energy particle-beam weaponry, it is his antigravity device that is not in use today. At least we don't think it is. And so, we believe, it was the same with Thomas Edison's spirit phone. That invention was so far ahead of its time and posed so much of a challenge to organized religion of the day that, even if only for the good of his companies, those running them after his death deemed it best not to release the information.

---

80 Birnes, William, "The Incredible Mac Magruder Story," *UFO Magazine*, 21 (4), (June, 2006), 32–39.

FAITH-BASED COMMUNICATION WITH THE DEPARTED

It is ironic that the major religions have a faith-based belief in the survival of the soul after the body dies. Christians may seek the help of saints, other religions may worship ancestors, still other forms of devotion may worship deities of nature. However, when it comes to making actual physical contact with the departed, that is normally condemned as heresy or the work of the devil. This, we believe, is why, whether or not Edison actually succeeded in getting a registration on his spirit phone, no physical records of that device exist. We may adhere to our faith and treasure our science, but, even in today's technological society, the two very rarely mix.

CHAPTER 19

The Legacy of Thomas Edison: Successes, Failures, Controversies, Meanings

Thomas Edison has reached the rare pinnacle of American saint-hood, having been so translated by science and literary historians whose hagiographies of the great inventor focus on his genius, foresight, and skill. But the version of Thomas Edison's life so promulgated is mostly based on a sanitized history of the man whose otherwise outlier beliefs are excised from the official story. His legacy is almost immeasurable because he literally created the communications-based society that we live in today. His inventions defined modern society, which, in turn, defines us. Therefore, his experimentation with a spirit phone, though it might seem enig-matic, was actually a logical extension of his attempt to push the envelope of reality to a new frontier.

Whether the spirit phone worked or not; whether the spirits of the departed are actually cohesive life units or not; whether, during that all-important first test at the beginning of the 1920s, spirits were actually summoned by mediums and clairvoyants and crossed the photon beam of Edison's device or not, none of this detracts from the lasting effects Edison had not only upon American society, but upon the world. Just in terms of his two major entertainment industry inventions, the phonograph and the motion-picture camera and projector, Edison's work turned America into a powerhouse of worldwide social culture. It still resonates today.

LEGACY OF THOMAS EDISON

Regardless of spirit phone proof of concept, the legacy of Thomas Edison has become an American legend, the story of our society as well as the personal triumph of Edison. For example, Edison's family's social ascendance was a quintessential American story about the rise of a family from the working class to industrial and governing elite. Sam Edison, a Canadian immigrant, was a hardscrabble shopkeeper and laborer. His son, Thomas, a scientist, invented the twentieth century, pointed America to the next century, and was one of his society's greatest industrialists. Sam Edison's grandson, Thomas's son Charles, became the governor of New Jersey, the Edisons' adopted state.

Thomas Edison's legacy involves the three century-defining and social-defining industries that he created: recording, motion pictures, communications, and municipal electrical power grid.

And his merging of the recording and picture and communications industries, distributing content along an ancillary grid as part of a municipal power grid, laid out the basis for today's Internet.

SCIENTIA VINCIT OMNIA

One of the lasting social effects that Thomas Edison left us was his belief that science can solve problems, even problems of the spirit. This was the essence of his spirit phone, to use science to prove that there is a reality to the spirit that's measurable and with which we can communicate. This was the holy grail of his, and Tesla's, final years. The fact that we have no records of Edison's device's success doesn't alter his belief that it was possible. At the least, Roger Penrose provides a quantum-based explanation of consciousness and the possibility that it lingers in an entangled quantum state after the body's death. We know that Edison had a deep belief in the power of science to find the truth about life after death. We know that he reiterated to his followers and to audiences that the soul is immortal and we do not die. Though a materialist, he believed that using science to reveal the unseen in our reality was doable.

Edison was also one of the most important harbingers of the age of modern industrialism. Although Age of Science and Industrialism itself began in the nineteenth century, by the early twentieth century there was a new burst of industrialism that embraced the science of mass production and manufacturing as well as the sciences of theoretical physics and medicine and the science of mass communication. In this new age of industrial science, Edison, Harvey Firestone, Henry Ford, David Sarnoff, John D. Rockefeller, Andrew Mellon, and Andrew Carnegie were all the industrial and financial barons who helped invent the century and create the environment that is still with us today. But it was Edison, only, who was the first to create an assembly line of science as opposed to an assembly line of product. Where Ford was able to marshal the skills of laborers to downstream the fabrication of automobiles, Edison took it one step further and worked from the stage of actual creation of the idea—the application of science to a market-fulfilling product—to the mechanism for mass producing the proof of concept. In this he was one of the most creative business personalities of his age.

In fact, even while he was alive, but certainly after he died, Thomas Edison was one of the few people who actually reached a form of American sainthood, if that is possible. He did this by surmounting his own humanity, his foibles and downright weaknesses, and through legend and lore was transmogrified.

GLORIFIED BY THE MOTION-PICTURE INDUSTRY

The very industry that Edison created, motion pictures, bestowed upon him its own version of sainthood with two biographical films. In 1940, nine years after Edison's death, Louis Mayer's MGM made two motion pictures about the great inventor, films that were headed up by the most valuable talent the studio had under contract at the time—Mickey Rooney in *Young Tom Edison* and Spencer Tracy in *Edison the Man*. These two films were hagiographic pieces, the first with Mickey Rooney playing the inquisitive youth always getting into trouble but ultimately always doing the right thing. The second looked at Edison in 1929, just two years away from his death, standing at a testimonial awards dinner in which he looks back on his life.

For the making of *Young Tom Edison*, it was necessary that the Edison home and laboratory in Port Huron, Michigan, be recreated from whatever could be relocated from Menlo Park and Fort Myers, where Edison had lived next to Henry Ford. Ford took the motion picture so seriously that he actually wanted the film to be shot in Edison's home of Port Huron and, to facilitate that, he arranged on his own dime to have the entire Edison laboratory relocated and restaged at Port Huron.[81] Afterwards, Ford was so impressed with Mickey Rooney's performance in the film that he

---

81 Lertzman, Richard A. and William J. Birnes, *The Life and Times of Mickey Rooney*, (New York: Gallery Books, 2015).

arranged for Rooney to meet with Edison's widow. Ford's friendship with Rooney continued well into the next decade, as did Edsel Ford's friendship with Mickey Rooney. In fact, Ford made sure that the very first Edsel that rolled out off the assembly line was delivered to Mickey Rooney, who bragged about it well into his eighties.

It took just two generations for the Edisons to become one of the most powerful families in America. And it took three generations for an Edison to rise to political power. This is how America works. But besides the financial story of the Edisons and their rise to prominence, it is the intellectual story of Thomas Edison that is so intriguing. He was more than just the right person at the right time. He actually invented the future.

HOW THE EVOLUTION OF EDISON'S INVENTIONS WORKED: THE ELECTRIC PENCIL

It is fascinating to track the evolution of Edison's ideas into products because it demonstrates his abilities to see beyond the immediate invention to industry that would result. One of the best examples of this is Edison's invention of the electric pencil, which led to the invention of the mimeograph machine, a device with which an operator would mark a letter or image on a stencil, which would then imprint on a piece of paper. This created the duplicating industry and brought the power of print, which in the 1870s was only a power available to industry because of the expense of heavy printing presses, to smaller businesses and schools. In this, Edison was like the inventor of moveable type, a nineteenth-century Gutenberg, who brought graphic duplication to the world. And how ironic it was that Edison's first business was based upon his acquisition of used moveable lead type for his Grand Trunk Railroad newspaper when he was still a boy. The modern mimeograph machine, popular as recently as the 1970s and then superseded by the photocopier, revolutionized the dissemination of information. Imagine a simple device for reproducing designs and text, improved over the decades, which lasted over one hundred years. This was Edison's vision.

THE R&D FACTORY

His other vitally important vision was the idea that inventions could be imagined and brought to fruition within a laboratory setting, corralling the talents of many different individuals with different skills. In this way, it wasn't just the invention itself that came out of Edison's Menlo Park factory, it was the creation of what today we call the research and development shop, a function so critical to industry that R&D is a line item on corporate budgets. Research and development is the heart and soul of industries that need to stay ahead of their competitors, that need to foresee the need for new products to provide solutions for things consumer and business markets might not even know exist, and that need to create value out of intellectual product.

THE QUEST FOR FIRE

One of the seminal moments in the development of human civilization was the utilization of fire. It brought light to the night, brought humans together to share the food they cooked, and provided warmth. Fire also brought about the rapid development of structured vocal human language, and from structured language to concepts, ideation, and literacy. Though this may seem like a stretch at first, the electric light and municipal power grid also extended the reach and growth of civilization. This wasn't just the creation of a new industry, public electric utilities; it enabled cities to grow, to electrify themselves to bring more light than natural gas provided, and to electrify mass transportation. Thus, the advancement of human civilization in the twentieth century can be attributed to Edison's experiments with electricity in the late nineteenth century. The invention of the light bulb, and thence the Edison tube, became the basis for communication via electrical circuitry, particularly television. There is also the socioeconomic legacy of Thomas Edison. Perhaps his biggest legacy was the consumerization of a market previously only open to the rich, appliances and convenience appliances could be produced for the masses instead of only for the upper classes. He brought about a greater egalitarian distribution of goods, which, in turn, brought about a greater spread of opportunities for advancement and growth for the working classes. In so doing, he transformed America in the twentieth century.

THE CREATION OF THE SCIENCE OF INVENTION

It can be said that the science of invention, if not the practice of science itself, is a form of intellectual reverse engineering. Scientists begin with a theory of something—say the theory that the lightning we see striking the earth is a form of electricity. Then, scientists, like Benjamin Franklin, use experimentation—say a lightning rod attached to a high-flying kite—and the design of what constitutes proof to certify that theory is either correct or incorrect. But in either instance, it starts with a theory, sometimes only based upon a hunch.

This reverse engineering of thought to practice to proof also best describes how Edison worked and why his work was so consumer oriented. It also explains the process by which he formulated his ideas for the spirit phone. Because he lived and worked in the age of Einstein and Planck and was a consumer of intellectual thought during the age of Freud and Jung, he was able to develop his theory of life units. But, he advanced the thought by supposing that if electronic particles had qualities that allowed for attraction, perhaps these life units also possessed the physical attribute of attraction. And, even half predicting the theory of quantum entanglement and spooky attraction, Edison theorized that these life units were able to corral themselves into specific cohesive masses.

Then, inasmuch as life units formed the basis for how human bodies navigated and rejuvenated themselves, he supposed, they not only pre-existed the body, but were the building blocks of life. And here he was advancing his own concepts that would become the theory of the genome and the work of DNA and RNA. But where he took it, no one had gone before. His thought process ultimately merged spiritualism and religion. Edison believed that we don't really die, and that this could be assayed scientifically. And that, as stressed throughout this book, was why the spirit phone would work.

We may take Edison's life and work for granted today, now that we're in the digital age in which human beings and machines are actually merging to evolve into a new lifeform, the cybernetic organism with paralyzed limbs directed by computer and brain implants to merge thought, volition, and machine, but it was the process of Edison's thought that brought us to this point. It's not just the inventions, but the intellectual process itself behind the inventions that brought them into existence. That is Edison's greatest legacy.

The mere possibility that science could prove that the unseen could be perceived as the driving force behind Edison's vision of the importance of electricity and the measurement thereof. It is, therefore, a more than fitting tribute that over sixty years after the death of Albert Einstein, the scientist who probably influenced

Edison the most, one of the major aspects of his theory of General Relativity was borne out. This occurred in 2016, when a gravitational wave ever so slightly altering the space/time of planet Earth reached the Laser Interferometer Gravitational-Wave Observatory, our intergalactic version of a nineteenth-century tromometer, after its 1.3-billion-mile journey across creation, where it was duly registered. Of course, this gravitational wave did not prove that Edison, or Tesla for that matter, could talk to the dead. But it did show that both Edison and Tesla were correct in their respective theoretical pursuits of a device that could measure what had only been dreamed of for scores of thousands of years: a device to ping the unknown.

Edison was devoted to the process of taking research into theory, then developing inventions based on the proof of that theory and tailoring them to a consumer market that could exist. In the final decade of his life, there was an insight into a market that not only existed at the time, but had existed since the beginning of civilization. This was a market that asked, and sought explanations for, the basic questions concerning what is life and what happens when life is over. Imagine a person who had invented the very things that electrified and mechanized an entire civilization, now in the final decade of his life, going back to the beginning of human existence at the mouths of caves to answer the primal questions that remained for the hundred millennia of human development. Was that because Edison knew that he would be staring into the face of death and believed that the only way to understand it was through science? Or maybe it was because the person who lit up the darkness of the urban night now wanted to light up the last realm of darkness that existed: the darkness of death. If science could create cities of light, why couldn't science light up death itself?

This was how the process of Edison's thought led him to the creation of the spirit phone. At the very last, when he awoke from his coma in 1931, he was finally able to answer the question that he had been working for ten years: the spirit lives on and we don't die.

About the Authors

Joel Martin is a national bestselling author, a former Long Island, New York, radio and television host who won a Cable ACE Award, and a nationally recognized expert on the paranormal. He was an investigative reporter who exposed the Amityville horror story as a hoax. Martin has co-authored with Birnes the *Haunting of America* trilogy and *The Haunting of the Presidents*. Joel Martin lives on Long Island.

Bill Birnes is the *New York Times*–bestselling author of *The Day After Roswell*, the consulting producer and lead host of History's *UFO Hunters*, and the guest expert on *Ancient Aliens*. Dr. Birnes has co-authored *Dr. Feelgood* with Rick Lertzman and *Hearts of Darkness* and *Wounded Minds* with Dr. John Liebert, all published with Skyhorse Publishing. Currently, Dr. Birnes is the Chairman of the Board at Sunrise Community Counseling Center in Los Angeles and a National Endowment for the Humanities fellow. Birnes lives in New Hope, Pennsylvania, with his wife, novelist Nancy Hayfield, who wrote *Cleaning House* and *Cheaper and Better*.

Bibliography

Adair, Gene. *Thomas Alva Edison: Inventing the Electric Age*. New York: Oxford University Press, 1996.

Anderson, Kelly. *Thomas Edison*. San Diego, CA: Lucent Books, 1994.

Baldwin, Neil. Edison: *Inventing the Century*. New York: Hyperion, 1995.

Barnham, Kay. *Thomas Edison*. Chicago: Raintree, 2014.

Brands, H. W. *The First American: The Life and Times of Benjamin Franklin*. New York: Doubleday, 2000.

Broughton, Richard S. *Parapsychology: The Controversial Science*. New York: Ballantine Books, 1991.

Cawthorne, Nigel. *Tesla vs. Edison: The Life-Long Feud That Electrified the World*. New York: Chartwell Books, 2016.

Cheney, Margaret. *Tesla, Man Out of Time*. Englewood Cliffs, N, 1981.

Cramer, Carol, Ed. *Thomas Edison*. San Diego, CA: Greenhaven Press, 2001.

Daly, Michael. Topsy: *The Startling Story of the Crooked-Tailed Elephant, P.T. Barnum, and the American Wizard, Thomas Edison*. New York: Atlantic Monthly Press, 2013.

D'souza, Dinesh. *Life After Death: The Evidence*. Washington, D: Regnery Publishing, 2009.

Friedel, Robert, Paul Israel, With Bernard S. Finn. *Edison's Electric Light: The Art of Invention*. Baltimore, MD: Johns Hopkins University Press, 2010.

Galvin, Anthony. *Old Sparky: The Electric Chair and the History of the Death Penalty*. New York: Carrel Books, 2015.

Hoffmann, Banesh with Helen Dukas. *Albert Einstein: Creator and Rebel*. New York: Viking Press, 1972.

Hunt, Inez And Wanetta Draper. *Lightning in His Hand: The Life Story of Nikola Tesla*. Denver, CO: Sage Books, 1964.

Israel, Paul. *Edison: A Life of Invention*. New York: John Wiley & Sons, 1998.

Johnston, B. (Ed.) *My Inventions: The Autobiography of Nikola Tesla*. Williston, VT: Hart Brothers, 1982.

Kubala, Thomas. *Electricity 1: Devices, Circuits, and Materials*. Clifton Park, New York: Thomson Delmar Learning, 2006.

_____. *Electricity 2: Devices, Circuits, and Materials*. Clifton Park, New York: Thomson Delmar Learning, 2006.

Keljik, Jeff. *Electricity 3: Power Generation and Delivery*. Clifton Park, New York: Thomson Delmar Learning, 2006.

Lewis, James R. *Encyclopedia of Afterlife Beliefs and Phenomena*. Detroit, MI: Gale Research, 1994.

Mason, Paul. *Thomas A. Edison*. Austin, TX: Raintree Steck-Vaugn, 2002.

Melton, J. Gordon, Ed. *Encyclopedia of Occultism & Parapsychology*. Detroit, MI: Gale Research, 2001.

Moore, Graham. *The Last Days of Night: (A Novel about Edison)*. New York: Random House, 2016.

O'Neill, John J. *Prodigal Genius: The Life of Nikola Tesla*. Kempton, IL: Adventures Unlimited Press, 2008.

Oxlade, Chris. *Using Electricity*. Chicago, IL: Heinemann Library, 2012.

Parker, Steve. *Fully Charged Electricity*. Chicago, IL: Heinemann Library, 2005.

Seifer, Marc. *Wizard: The Life and Times of Nikola Tesla: Biography of a Genius*. Secaucus, NJ: Citadel Press, 1996.

Simon, Linda. *Dark Light: Electricity and Anxiety From Telegraphy to the X-Ray*. New York: Harcourt, 2004.

Time-Life Books: *This Fabulous Century: 1910–1920*.

_____: *This Fabulous Century: 1920–1930*.

Uschan, Michael. *The 1910s*. San Diego, CA: Lucent Books, 1999.

Wallace, Joseph. *The Lightbulb*. New York: Atheneum Books for Young Readers, 1999.

Woodside, Martin. *Thomas A. Edison: The Man Who Lit the World*. New York: Sterling, 2007.

OTHER RESOURCES: LIBRARIES AND COLLECTIONS

Milan Public Library, Milan, Ohio

Edison Birthplace Museum, Milan, Ohio

Thomas Edison Center/Menlo Park Museum, Edison, New Jersey

Thomas A. Edison Papers, Rutgers, The State University Of New Jersey, Piscataway, New Jersey

Thomas Edison National Historical Park, West Orange, New Jersey

Greenfield Village (Edison at Work Historic District), Dearborn, Michigan

Tesla Science Center at Wardenclyff, Shoreham, New York

Liberty Science Center, Jersey City, New Jersey

Chicago Museum of Science and Industry

New York Public Library

East Meadow Public Library, East Meadow, New York

New York Times

Scientific American Magazine

Century Magazine (April 1895)

New York Sun (1896)

New York Herald (1897)

Index